AF843938

The Future of Food Policy in the UK

This book makes a compelling case for the urgent need to develop a new approach to food and farming—one that integrates the principles of food security and sovereignty.

In the face of escalating challenges such as climate change and increasing political instability worldwide, it is essential to rethink food production and future-proof it by prioritising resilience. While food security and sovereignty are widely recognised and debated concepts, this book considers the critical yet often overlooked dimension of food stability as a cornerstone for shaping future food policy. As public interest in sustainable food systems continues to grow, this book provides readers with a deeper understanding of the complexities of the food system and offers actionable insights into how consumption patterns can be shifted to foster a more sustainable future. Tracing the journey from farm to fork, the chapters explore key topics such as climate resilience, genetic diversity, supply chains, food resilience, and market competition. With a particular focus on the UK food system, this book examines the implications of the 2025 UK Government Food Strategy for England and its potential to reshape the future of food governance.

This book is an essential resource for anyone seeking to better understand UK food policy and the pathways to achieving a secure, equitable, and sustainable food future.

Honor May Eldridge has worked on sustainable food policy for the past decade in both the USA and the UK. During this time, she has focused on the environmental and social impact of agricultural systems and produced a wealth of reports and briefings for political and civil society audiences. She has worked both in and out of government, including for the Food Standards Agency, Sustainable Food Trust, the Soil Association, and the Center for Food Safety. She is the author of *The Avocado Debate* (Routledge, 2023) and *The Future of Food: Issues and Innovations in the 21st Century and Beyond* (2026).

Routledge Studies in Food, Society and the Environment

The Real Cost of Cheap Food
Michael Carolan

Plant-Forward Cuisine
Basic Concepts and Practical Applications
Ole G. Mouritsen, Klavs Styrbæk and Mariela Johansen

Ethical Omnivores
Better Eating for Everyone
Samantha Noll

Urban Food Systems in Latin America
Territories, Mobilities and Governance
Edited by Tiana Bakić Hayden and Joaquín Pérez Martín

International Causes of Hunger and Malnutrition
Food Insecurity and the Global Economy
Francis Adams

UK Food Policy History
Economic, Social and Political Influences
Martin Caraher and Sinéad Furey

Sustainable Consumption and Everyday Food Practices in Europe
Edited by Virginie Amilien, Monia Saïdi and Matthieu Duboys de Labarre

The Future of Food Policy in the UK
Recipe for Change
Honor May Eldridge

For more information about this series, please visit: www.routledge.com/Routledge-Studies-in-Food-Society-and-the-Environment/book-series/RSFSE

The Future of Food Policy in the UK

Recipe for Change

Honor May Eldridge

Routledge
Taylor & Francis Group
LONDON AND NEW YORK

earthscan
from Routledge

First published 2026
by Routledge
4 Park Square, Milton Park, Abingdon, Oxon OX14 4RN

and by Routledge
605 Third Avenue, New York, NY 10158

Routledge is an imprint of the Taylor & Francis Group, an informa business

© 2026 Honor May Eldridge

The right of Honor May Eldridge to be identified as author of this work has been asserted in accordance with sections 77 and 78 of the Copyright, Designs and Patents Act 1988.

All rights reserved. No part of this book may be reprinted or reproduced or utilised in any form or by any electronic, mechanical, or other means, now known or hereafter invented, including photocopying and recording, or in any information storage or retrieval system, without permission in writing from the publishers.

For Product Safety Concerns and Information please contact our EU representative GPSR@taylorandfrancis.com. Taylor & Francis Verlag GmbH, Kaufingerstraße 24, 80331 München, Germany.

Trademark notice: Product or corporate names may be trademarks or registered trademarks, and are used only for identification and explanation without intent to infringe.

British Library Cataloguing-in-Publication Data
A catalogue record for this book is available from the British Library

Library of Congress Cataloging-in-Publication Data
Names: Eldridge, Honor May author
Title: The future of food policy in the UK : recipe for change /
Honor May Eldridge.
Description: New York, NY : Routledge, 2026. |
Series: Routledge studies in food society and the environment | Includes bibliographical references and index.
Identifiers: LCCN 2026009047 (print) | LCCN 2026009048 (ebook) |
ISBN 9781032948294 hardback | ISBN 9781032948300 paperback |
ISBN 9781003581871 ebook
Subjects: LCSH: Nutrition policy—Great Britain | Food security—Great Britain
Classification: LCC TX360.G7 C47 2026 (print) | LCC TX360.G7 (ebook)
LC record available at https://lccn.loc.gov/2026009047
LC ebook record available at https://lccn.loc.gov/2026009048

ISBN: 978-1-032-94829-4 (hbk)
ISBN: 978-1-032-94830-0 (pbk)
ISBN: 978-1-003-58187-1 (ebk)

DOI: 10.4324/9781003581871

Typeset in Times New Roman
by codeMantra

This book is dedicated to the many hours that I have debated politics with my husband, and to the food future that my son and daughter will inherit.

Contents

Preface

This book sets out a bold vision for the future of food and farming—one rooted in the belief that we must move beyond outdated binaries, and build a truly resilient food system. For too long, the policy conversation has been split between food security and food sovereignty, as though these were opposing goals. But this division has hindered progress. It's not a matter of choosing one over the other; both are essential. Yet, even together, they leave a critical gap. That missing element is food resilience. It is the linchpin we need in a world facing escalating climate chaos and deepening geopolitical uncertainty. No longer a luxury, resilience is a necessity. Without it, neither food security nor sovereignty can be sustained over time.

Public interest in the sustainability of food has never been higher. Citizens are not only more aware of the environmental impacts of their consumption but are actively changing their habits. A 2021 Deloitte survey revealed that nearly half the UK public had shifted towards more local and seasonal produce, and almost a third had reduced meat and dairy consumption.[1] This shift reflects a growing desire to engage with food not just as consumers, but as citizens concerned with climate, health, and justice. Citizens play an important—though often constrained—role in shaping a resilient food system. Every purchase influences the kind of agriculture, trade, and environmental impact the system sustains. Yet individual choice operates within a curated food environment dominated by supermarket supply chains, advertising, and pricing structures that limit agency and make sustainable options less accessible. Despite these constraints, citizens can make meaningful contributions by supporting systemic change, signalling to retailers and policymakers alike that fairness, transparency, and sustainability are public priorities.

With the UK entering a new political era following Labour's decisive victory in July 2024, there is a rare window of opportunity to rethink how we produce, distribute, and govern food. Conversations in Whitehall, particularly within the Department for Environment, Food, and Rural Affairs (Defra), have increasingly pointed towards the need for greater resilience that focuses on delivering food that is sustainable, affordable, and healthy for the public. This book speaks directly to this moment. It offers a timely contribution to a conversation already in motion and aims to shape the very framework of future legislation. Though centred on the UK, this book has broader relevance. In the wake of Brexit's economic and political upheaval, Britain stands at a unique crossroads. The National Food Strategy of 2025 has the potential to spark a profound shift in the understanding of food and farming—one that could inspire Global North nations to rethink their own approaches to food and farming. By presenting a clear, compelling case for a resilient food future—anchored in security, and sovereignty—this book offers both a roadmap and a rallying cry for a better way forward.

Note

1 Fenech, C., Walton, B., Majury, A., *The sustainable consumer. Understanding consumer attitudes to sustainability and sustainable behaviours.* Deloitte, 2024, https://www.deloitte.com/uk/en/Industries/consumer/perspectives/the-sustainable-consumer.html

Introduction

A fork in the road

How do we feed a nation in the 21st century? How do we create a food system that is fit for purpose and able to withstand the shocks that it will inevitably face? Societies across history have worried about how to feed themselves, knowing that ensuring an adequate and consistent supply of food is the most critical component for populations to thrive. Without it, economies cannot grow and the public becomes restless. Riots quickly and easily erupt when a population gets hungry. The UK is no different. Through history, we have faced pressure points when food supply was dwindling, costs were rising and the public felt hungry—and angry. For example, the 1766 Wheat Riots, where urban populations objected to the price of flour, or the 1917 protests in Glasgow over rationing and food shortages. This is not an issue of the distant past. More recently, a dispute between two shopkeepers over the sale of bread sparked riots on an estate on the edge of Cardiff in 1991.[1] Yet, the situation is more precarious today than then. It is true that global food supply has increased dramatically over the last century (grain production has increased from 692 million tonnes in 1950 to 2,849 million tonnes in 2024), and we now have enough food to feed the world.[2,3] But, the system of supply is unstable. We have shifted to a model of food security that is dependent on imports from around the world, which are vulnerable to disruptions—particularly from geopolitical turmoil and conflict. We have an agricultural system that has been industrialised, and while it churns out substantial quantities of crops, it is damaging our planet and destroying natural ecosystems on which we depend. The food it produces is also overarchingly unhealthy and is driving a crisis in global diets and public health.

We need to change. We need a new system that can deliver sufficient food for everyone equitably and can also allow our landscapes to heal and regenerate. Beyond that, however, we need a new approach to food and farming that recognises the vulnerability of the world that we face. We are entering a tumultuous period with geopolitical instability causing almost continuous shocks to the market. Trade wars, armed conflicts, and isolationist politics have all exposed how perilous our current system is. That is against a background of agriculture struggling to cope with the Climate and Nature Crises

DOI: 10.4324/9781003581871-1

that are only just beginning to show their devastating impacts. Climate change is amplifying risks, with extreme weather events, droughts, and floods disrupting production. Soil degradation, loss of biodiversity, and water scarcity further threaten long-term sustainability. Furthermore, growing economic inequality and corporate consolidation leave many small-scale farmers struggling with many exiting the industry, leaving us overly dependent on fewer and fewer large corporations. This system, designed for efficiency and profit, lacks resilience in the face of compounding crises—making it ill-equipped to meet the needs of a changing and uncertain world.

In the UK, we have a key opportunity to rethink our food and farming policies and create something bold and dynamic that recognises the challenges that we face and addresses them head-on. The UK's decision to leave the European Union following the 2016 referendum created the space to create a new agriculture policy. Until that point, UK farming policy had functioned under the umbrella of the EU's Common Agricultural Policy (CAP)—its framework for supporting agriculture and rural development across member states. Established in 1962, CAP has historically provided subsidies tied to production. While over the years, it has made reforms and shifted focus towards environmental stewardship, rural resilience, and climate action, the overarching focus has remained on production. After all, it is cut fairly hard to turn a tanker once it is in motion. Attempts to 'green' CAP have been bolt-ons—and fundamentally unable to change the core of a programme that is designed to encourage farmers to produce as much as possible. Now, outside of the EU and outside of CAP, the UK has an opportunity to design its own food and farming policies that aren't burdened by the past. Remainer or Brexiteer, the decision to leave the EU afforded the UK the enviable opportunity to create a food and farming policy from scratch, one that is reflective of where we are today. It could be designed to consider the realities of globalisation, climate change, and population growth. This moment was unique since every other Global North country has only been able to incrementally adjust its food policy while fighting the status quo. Instead, the UK could look to the future. The UK Government could create a food and farming policy framework that is fit for the next hundred years. One that addresses the challenges that we face as a nation and as a planet. One that ensures healthy and sustainable food for everyone, supports farmers and rural communities, and builds resilience into every step of our food system. By doing so, we can create a food system that can not only survive but also thrive.

Brexit, Health and Harmony

In June 2016, the UK voted to leave the EU in a referendum. Many of those who voted to leave highlighted the desire to take back control from the EU and make more legislative decisions at home in the UK. Rural voters (many, but not all, farmers) voted 55% to leave, and 83% of rural counties supported

Leave over Remain.[4,5] In the wake of the vote, there was a protracted and complex negotiation to decide the new relationship with the EU, including terms for trade. The ambiguity over the future trade dynamic led to significant political turmoil in the UK. Firstly, Prime Minister David Cameron resigned in the hours after the result came in. Theresa May took office but eventually fell from power in May 2019 after being deemed to be too 'soft' on the EU in negotiations.[6] She was replaced by Boris Johnson, who was considered to be a Brexit hardliner due to his role in the referendum campaign.

Outside the EU, the UK faced the momentous task of creating a new food and farming policy. The Conservative Government, under Theresa May, appointed Michael Gove to be the Secretary of State for the Department of Environment, Food, and Rural Affairs (Defra). This was considered by many to be a good sign of the Government's commitment to farming given his political stature. To a certain extent, they were right. Under Gove's leadership, Defra launched the Health and Harmony consultation and commissioned leading food expert, Henry Dimbleby, to write a National Food Strategy (known as the Independent Dimbleby-led Food Strategy) to which we will return later.[7,8] Published by the UK Government in 2018, Health and Harmony outlined a post-Brexit vision for agriculture, replacing the EU's CAP. It proposed a fundamental shift: rewarding farmers for delivering "public goods", such as clean air, thriving wildlife, improved soil health, and access to nature, rather than payments based on land ownership. The report emphasised environmental sustainability, animal welfare, and the long-term restoration of the countryside. It marked a significant moment in reshaping farming policy, aiming to align agricultural support with ecological outcomes and public benefit in a post-EU future.

It seemed a promising start, but then, in 2019, Gove left Defra for pastures new, and the spotlight went with him. Resources and ambition dwindled. Defra launched the Environmental Land Management (ELM) scheme in 2021.[9] Designed to replace ELM proposed rewarding farmers and land managers for delivering public goods and consists of three tiers: the Sustainable Farming Incentive (SFI), Local Nature Recovery, and Landscape Recovery. ELM represents a shift from area-based subsidies to outcome-based payments, aiming to support more sustainable, environmentally focused farming practices. While ambitious in scope, the rollout has faced criticism over delays, complexity, and concerns about financial viability for farmers during the transition period. While they had the opportunity to be bold, so far, ELM has failed to deliver the necessary catalyst for change and has perpetuated the siloed thinking of food and nutrition in one department and farming in another.

COVID

Yet, there was a moment when the Conservative Government (and the nation writ large) recognised the weaknesses of our food system and made efforts to correct them. COVID-19 upended the world and shook our entire system. The World Health Organization has estimated that nearly 15 million people *died* as a result of the pandemic, including both direct and indirect causes—a shocking number that fails to acknowledge the pain and loss experienced by so many.[10] With such a devastating event, the shock waves were extensive, and the food sector was badly impacted too. Supply chains faltered in the pandemic. Lockdowns, labour shortages, and transport restrictions led to delays in production, processing, and distribution. Panic buying and stockpiling worsened shortages in stores, and vulnerable communities faced increased food insecurity as access and affordability declined. The crisis exposed weaknesses in the global food system and emphasised the need for resilient, localised supply chains to better withstand future disruptions. Stay-in-place orders shuttered many ports and shipping routes, especially in the early days of the outbreak. Due to the globalised nature of our food supply, differing public health responses caused confusion and delay. For example, the UK imports $711 million AUD/£339 million goods annually from Australia. But, in response to COVID-19, Australia closed its international borders for nearly 2 years, imposed strict lockdowns (particularly in cities like Melbourne), and enforced quarantine measures for both travellers and workers. These measures, especially early on, created logistical and operational hurdles. Similarly, farmers in the UK struggled with workforce gaps, especially in harvesting and packaging, while perishable goods went to waste. Without the freedom of movement, many temporary seasonal workers were not able to get to their workplaces. Many workers were off sick with COVID. Some were simply too fearful of getting sick to travel or, understandably, wanted to remain close to family in a period of such turmoil.

The disruption exposed the vulnerability of the just-in-time system of retailers. Supermarket shelves are stocked with precision, not surplus. Goods arrive exactly when needed, timed perfectly to match predicted customer demand, so that there's little on-site storage. Retailers carefully monitor purchasing data to ensure that the right food is flowing to the shelves as and when it is needed so as to satisfy the customer without leaving stock on the shelves. However, one shock and the system falters. There is no margin for error, and there is very little hedging built into the model. COVID-19 was a massive shock, and customers routinely cleared grocery shelves—particularly in the early weeks of the lockdown, exacerbated by panic buying. This just-in-time approach is mirrored throughout the entire supply chain. The supply of food is continuous, with little kept in reserve, with little food kept in the country in reserve at any one time. The goal is efficiency: less waste, lower costs, fresher food. But it's a fragile dance. However, it is worth acknowledging how quickly this situation was righted, and the supply of food was rapidly restored.

New opportunities for change

July 2024 saw the arrival of a new Government in the UK. Labour won a landslide victory, with Sir Keir Starmer becoming Prime Minister. They swept the election with a promise of creating a new, more equitable, and just Britain. They set out a bold and pragmatic vision to rebuild the UK. Centred on five key missions—driving economic growth, delivering clean energy, restoring safety, expanding opportunity, and revitalising the NHS—their plan focused on long-term, practical change. Starmer appointed Steve Reed as the Secretary of State for Environment, Food and Rural Affairs, and Daniel Zeichner was appointed as Minister of State for Food Security and Rural Affairs. With this mandate, the Government has an opportunity to reconsider food and farming policy and create a holistic and interconnected approach to food that balances production and consumption in England to create a thriving farming sector that manages the land in an environmentally sensitive way, while simultaneously providing affordable and healthy food for all citizens.

When Labour came to power, the state of UK food and farming was precarious. UK agriculture and fishing sector was worth £13.1bn with food and drink manufacturing worth an additional £35.1bn, employing a total of 4.17 million people.[11,12] We export our food proudly across the world. Our main agricultural export is cereal (£2.4b per year), followed by beef (£2.1b) and dairy and eggs (£2.1b per year).[13,14] This export market rested on the long and proud history of UK food production that is associated with some leading food products, valued the world over. For example, Welsh lamb is prized in the Gulf States, and Cheddar cheese has a global reputation. However, underneath the surface, the picture did not look as rosy. By 2020, following years of consolidation, the UK has less than 30,000 small and medium farms and was hugely dependent on imports.[15] The UK currently imports 46% of its food, a total of £48bn annually, importing 75.6% from the EU, the largest trading partner for food—although this is admittedly down from 34.6 before the UK voted to leave the Union.[16] Our other main trading partners are the Netherlands (12.1%) and France (9.4%), although it is worth mentioning that the Netherlands, with its major port in Rotterdam, acts as a middleman, shipping food onwards across Europe and the world.[17]

Yet, despite that bountiful volume of food available, consumption isn't equitable. As of early 2025, approximately 7.3 million adults—about 14% of UK households—are experiencing food insecurity.[18] This includes 5% of households that reported going without food for an entire day.[19] Children are especially vulnerable, and shockingly, one in five children in the UK today is experiencing hunger. Sadly, this trend has been going in the wrong direction, and the situation has been slowly worsening over recent decades. The Government "austerity" programme, starting in response to the 2008 financial crisis, cut social provisions for many low-income families, such as Child and Housing

Benefits. With limited funding available, local councils were forced to cut back on provisions such as free school meals and community kitchens. This drove up hunger levels across the UK. The Trussell Trust reported handing out around 41,000 parcels in 2009–10, but by 2024, that number had grown to over 3 million parcels annually, with a clear surge following major austerity policies.

COVID-19 compounded UK hunger. Tough as it was under the Government's programme of austerity, the situation became even worse with the pandemic raging. People faced increased financial instability and struggled to access food, especially with the growing inflationary pressure pushing up prices. This led to a public campaign by footballer Marcus Rashford to shame Boris Johnson's Government into delivering free school meals during the holiday period in 2020 at the height of COVID.[20] The extended period of financial challenge and post-COVID fallout has continued this trend. There has been a 94% increase in food bank usage over the past 5 years.[21] As of June 2024, 2.7 million children across the UK experience hunger.[22] Labour committed to challenging these tensions when it came into office. In their 2024 Manifesto, they pledged to "end mass dependence on emergency food parcels", describing the current reliance on food banks as "a moral scar on our society".[23]

Yet so far, it has not quite gone to plan.... The Labour Government has done little to endear itself to farmers. One of their early decisions was to revoke the carve-out that agricultural land has under inheritance tax. In the 2024 UK Budget, UK Chancellor Rachel Reeves announced that, effective from April 2026, inherited agricultural assets exceeding £1 million will be subject to a 20% inheritance tax.[24] Previously, agricultural property was exempt from inheritance tax, but under the new policy, farmers would now be charged, although admittedly at half the standard rate of 40%. The justification was that the majority of small, family farms sat under this threshold. This angered farmers and resulted in significant protests, including by famed farmer, Jeremy Clarkson. Many thought it reflected a misunderstanding of the UK's farming sector and that the decision would weaken national food security and rural communities. To add insult to injury, Defra announced a sudden closure of new applications for the Sustainable Farming Incentive in March 2025 without any warning.[25] Many farmers who had not yet submitted their SFI proposal were caught short. Zeichner suggested that this was a sign of the scheme's success that it was over-subscribed, as opposed to a sign that farmers felt high levels of income vulnerability.[26] Zeichner defended the decision, saying that the budget for the scheme had now been spent and there was no further money to distribute.[27] He also announced an intention to revisit the SFI to design it to make it work better for UK farmers.[28] The UK Government need to rebuild the trust with farmers that they lost through these actions. Farmers are essential stakeholders and need to become partners in delivering the future of our food policy—without farmers at the table, we cannot begin to tackle the challenges that we face. If the Government is going to reconceptualise SFI as it promised, there is an opportunity to connect farming with food and demonstrate the Government's commitment to farmers in the process.

National Food Strategy

During his tenure at Defra, Secretary of State Michael Gove commissioned an independent National Food Strategy.[29] To ensure its effectiveness, Gove asked Henry Dimbleby to lead it in 2019. Dimbleby was a good choice. He is a British businessman, cookery writer, and prominent advocate for sustainable food systems. He co-founded the fast-casual restaurant chain Leon in 2004, aiming to offer "natural fast food", and in 2009, he co-founded the Sustainable Restaurant Association, an organisation dedicated to promoting sustainability within the food service industry. The independent Dimbleby-led National Food Strategy (also referred to as the Dimbleby Strategy) was published in two parts. Part One was released on 29 July 2020 in response to the devastating impacts of COVID on the food supply and farming sector.[30] Part Two was published on 15 July 2021 and presented a more comprehensive set of proposals.[31] This document reviewed the UK food system and made clear recommendations as to how it could be improved to better deliver a sustainable, healthy and equitable future for UK food. It called for a sugar and salt reformulation tax to drive healthier products, and the expansion of free school meals and Healthy Start to improve children's diets. To tackle inequality, it urged stronger support for low-income families' access to good food. The report also emphasised investment in agricultural innovation and nature-friendly farming to protect biodiversity and reduce emissions. Central to its vision was a long-term national food plan. Without systemic reform, it warned, the UK faces worsening health crises and accelerating environmental collapse.

The Government responded by announcing that it would release its own Food Strategy in response, which was published in July 2025.[32,33] Given that there had been a change in Government since the release of the Dimbleby Strategy, it presented a key opportunity for Labour to set out its priorities for the food system now that it was in power and connect it to its wider agenda. It sets out a vision for a food system that is healthier, more sustainable, and supportive of economic growth. It acknowledges food's deep connections to health, the environment, culture, and the economy, while highlighting how current patterns drive inequality, ill health, and environmental damage. It highlights the need to protect the food system against climate change, market volatility, and geopolitical disruptions, while calling for greater domestic production of fruit and vegetables, reduced dependence on vulnerable supply chains, and more diverse and sustainable farming systems. Against this backdrop, the strategy seeks to promote "good food cycles". As it explains: "good food cycles are when healthy, sustainable options become the default – widely available, affordable, and appealing, supported by supply chains and marketing that reinforce healthier choices".[34] Such cycles create positive feedback, where healthier demand drives supply, investment, and innovation, reinforcing healthier and more sustainable consumption.

It is worth noting that both the 2025 Strategy and the Dimbleby Strategy were built on the Food 2030 strategy, which was an overarching government policy published in 2008 to advance sustainable and healthy diets.[35] Food 2030 puts forward a proposal for how the UK can maintain food security and ensure its citizens are able to eat properly in the face of rising global population, environmental threats, and increases in health issues like obesity. It does not propose specific policies, but lays out guidelines, including enabling and encouraging a healthy diet; ensuring a resilient, profitable, and comprehensive food system; increasing food production sustainably; reducing the food system's greenhouse gas emissions; reducing, reusing, and reprocessing waste; and increasing the impact of technology. It also encourages sustainable farming and trading with global partners.

Changes afoot

In September 2025, Prime Minister Starmer launched a reshuffle of his cabinet that brought major changes to Defra.[36] Prime Minister Keir Starmer removed Steve Reed from his post as Environment Secretary, ending his short tenure in the role. In his place, Emma Reynolds was appointed to lead the department, a move widely interpreted as an effort to strengthen the government's green and rural agenda while maintaining credibility on environmental issues. Daniel Zeichner, who had served as Minister of State for Food Security and Rural Affairs at Defra since July 2024, was removed from his post in the September 2025 cabinet reshuffle and returned to the backbenches. He was succeeded by Dame Angela Eagle, who assumed the role of Farming Minister within Defra alongside Emma Reynolds's appointment as Secretary of State.[37,38] The reshuffle reflects Starmer's broader effort to consolidate his cabinet team after the resignation of Deputy Prime Minister Angela Rayner triggered widespread changes. For Defra, the leadership shift could bring a sharper focus on food security, land management, and the transition to more sustainable farming systems. How Reynolds manages these priorities will shape the government's credibility on rural and environmental policy in the years ahead.

These changes all come against a backdrop of a world economy in turmoil, where the full impacts are still unfolding. Upon returning to office for a second term, President Trump reignited his protectionist agenda, positioning tariffs at the heart of his "America First" strategy.[39] Early in 2025, he targeted Mexico and Canada with new duties, before escalating dramatically in April with a sweeping package of global tariffs—including a 10% levy on UK goods.[40,41] Trump justified these actions as a way to restore U.S.

manufacturing, strengthen jobs, and rebalance what he deemed unfair trade relationships, especially with China.[42] Acknowledging there might be "some pain" for consumers, he argued on Truth Social that this was a price worth paying.[43] The immediate fallout was swift: global markets tumbled, trade flows were disrupted, and economists warned of sharp inflation as higher costs were inevitably passed down supply chains.[44] Although Trump temporarily froze tariffs for 90 days following a significant U.S. stock market slide, the episode underscored the fragility of an interconnected global economy.[45] For food systems, the risks are particularly acute. With supply chains already stretched by climate shocks, war, and energy volatility, populist protectionism of this kind threatens to drive food prices higher, destabilise farmers' livelihoods, and reduce access to affordable food worldwide. In a globalised food system, unilateral trade wars ripple far beyond national borders—undermining resilience at the very moment when cooperation is most urgently needed.

Consequently, the time to refocus our food system on resilience is now. We are facing interconnected Climate and Dietary Crises that will have increasingly dramatic impacts, and when coupled with global population growth and geopolitical instability, the potential disruptions to our food supply are profound. Yet within this challenge lies a political opportunity: the new Labour Defra team has a chance to reshape the future of food in the UK through the delivery of the National Food Strategy by rooting actions in a vision of a resilient, sustainable, and equitable farming system. Coming to power in the post-Brexit era, they can finally seize the opportunity squandered by the last Government to design a support framework outside the confines of the CAP—one that moves beyond the post-war fixation on yield and productivity and instead responds to the food challenges of the 21st century so that we can create a new dynamic that drives change and delivers a better food future for the next generation.

Notes

1 Evans, Max, Ely bread riots; painful memories linger 25 years on, *BBC News*, 2016, https://www.bbc.co.uk/news/uk-wales-south-east-wales-37183995

2 Easterbrook, Greggs, Forgotten benefactor of humanity, Atlantic, 1997, https://www.theatlantic.com/past/docs/issues/97jan/borlaug/borlaug.htm

3 World Hunger Facts, Action against hunger, 2020, https://www.actionagainsthunger.org.uk/why-hunger/world-hunger-facts

4 Wilson, Brian, A rural analysis of the EU referendum results, the rural service network, 2016, https://www.rsnonline.org.uk/images/APPG/minutes/brexit/Rural_Analysis_of_EU_Referendum_Result.pdf

5 Uberoi, Elise, Briefing: European Union Referendum, 2016, House of Commons Library, 2016, https://researchbriefings.files.parliament.uk/documents/CBP-7639/CBP-7639.pdf

6 McTague, T. & Cooper, C., Inside Theresa May's Great British Failure, Politico, 2019, https://www.politico.eu/article/theresa-may-brexit-referendum/

7 UK Gov, Health and harmony: The future for food, farming and the environment in a Green Brexit – Policy statement, DEFRA, 2020, https://www.gov.uk/

government/publications/the-future-for-food-farming-and-the-environment-policy-statement-2018/health-and-harmony-the-future-for-food-farming-and-the-environment-in-a-green-brexit-policy-statement

8 UK Gov, National Food Strategy: The plan, DEFRA, 2021, https://www.nationalfoodstrategy.org/wp-content/uploads/2021/10/25585_1669_NFS_The_Plan_July21_S12_New-1.pdf

9 Scott, Edward, In focus: Environmental land management: Recent changes to the sustainable farming incentive and countryside stewardship scheme, House of Lords Library, 2024, https://lordslibrary.parliament.uk/environmental-land-management-recent-changes-to-the-sustainable-farming-incentive-and-countryside-stewardship-schemes/

10 Rosengren, Helen, 14.9 million excess deaths associated with the COVID 19 pandemic in 2020 and 2021, WHO, 2022, https://www.who.int/news/item/05-05-2022-14.9-million-excess-deaths-were-associated-with-the-covid-19-pandemic-in-2020-and-2021

11 UK Gov, Farming evidence pack: A high-level overview of the UK agricultural industry, DEFRA, Government Statistical Services, 2024, https://assets.publishing.service.gov.uk/media/68fb226d30c331c88be6f061/Farming_evidence_pack_2025_20251023.pdf

12 UK Gov, Food statistics in your pocket, DEFRA, 2023, https://www.gov.uk/government/statistics/food-statistics-pocketbook/food-statistics-in-your-pocket#gross-value-added-of-the-uk-agri-food-chain-2022

13 Scotch whiskey industry records 5.4b global exports in 2024 amid turbulent global trading conditions, Scotch Whiskey Association, 2024, https://www.scotch-whisky.org.uk/newsroom/2024-export-figures/

14 GOV.UK, Chapter 13: Overseas trade, DEFRA, 2024, https://www.gov.uk/government/statistics/agriculture-in-the-united-kingdom-2024/chapter-13-overseas-trade

15 Allen, Abby, The slow disappearance of small family farms, Pipers Farm, 2021, https://pipersfarm.com/blogs/journal/the-slow-disappearance-of-small-british-family-farms

16 GOV.UK, Chapter 13: Overseas trade, DEFRA, 2024, https://www.gov.uk/government/statistics/agriculture-in-the-united-kingdom-2024/chapter-13-overseas-trade

17 GOV.UK, Chapter 13: Overseas trade, DEFRA, 2024, https://www.gov.uk/government/statistics/agriculture-in-the-united-kingdom-2024/chapter-13-overseas-trade

18 Francis-Divine, Brigid, Maleek, Xammerah, & Foley, Niamh, Research briefing: Food poverty: Household, food banks and free school meals, House of Commons Library, 2025,https://commonslibrary.parliament.uk/research-briefings/cbp-9209/

19 Food Insecurity Tracking, The Food Foundation, 2020, https://foodfoundation.org.uk/initiatives/food-insecurity-tracking#tabs/Round-16

20 Siddique, Haroon, Marcus Rashford forces Boris Johnson into U turn on child food Poverty, *Guardian*, 2020, https://www.theguardian.com/education/2020/nov/08/marcus-rashford-forces-boris-johnson-into-second-u-turn-on-child-food-poverty

21 End of year food bank stats, Trussel Trust, 2024, https://www.trussell.org.uk/news-and-research/latest-stats/end-of-year-stats

22 Latest food insecurity tracker shows seven million adults going hungry, The Food Foundation, 2025, https://foodfoundation.org.uk/news/latest-food-insecurity-tracker-shows-seven-million-adults-going-hungry#:~:text=Key%20findings%3A,or%20get%20access%20to%20food

23 Change: Labour Party manifesto, Labour Party, 2024, https://labour.org.uk/wp-content/uploads/2024/06/Change-Labour-Party-Manifesto-2024-large-print.pdf
24 Geiger, Chas, Most farmers won't pay inheritance tax, PM insists, *BBC*, 2024, https://www.bbc.co.uk/news/articles/c4gv8zwge10o
25 Case, Philip, Calls for DEFRA budget transparency after SFI closure, *Farmers Weekly*, 2025, https://www.fwi.co.uk/news/farm-policy/calls-for-defra-budget-transparency-after-sfi-closure
26 Zeichner, Daniel, Sustainable farming incentive, Hansard, UK Parliament, 2025, https://hansard.parliament.uk/Commons/2025-03-12/debates/45069333-2542-4FFC-9A87-C4529D99209C/SustainableFarmingIncentive
27 Zeichner, Daniel, Sustainable farming incentive, Hansard, UK Parliament, 2025, https://hansard.parliament.uk/Commons/2025-03-12/debates/45069333-2542-4FFC-9A87-C4529D99209C/SustainableFarmingIncentive
28 Brown, Rachel, Defra stops accepting new applications for SFI 2025, *Farmers Guardian*, 2025, https://www.farmersguardian.com/news/4410695/defra-stops-accepting-applications-sfi
29 Gov.uk, Gove appoints Dimbley to pioneer 'farm to fork' revolution, DEFRA, 2019, https://www.gov.uk/government/news/gove-appoints-dimbleby-to-pioneer-farm-to-fork-revolution
30 UK Gov, Part one – National Food Strategy, DEFRA, 2020, https://www.nationalfoodstrategy.org/part-one/
31 UK Gov, National Food Strategy: The plan, DEFRA, 2021, https://www.nationalfoodstrategy.org/wp-content/uploads/2021/10/25585_1669_NFS_The_Plan_July21_S12_New-1.pdf
32 Lyne, Rachel, The UK announces collaborative development of new national food strategy, Browne Jacobson, 2025, https://www.brownejacobson.com/insights/food-and-drink-update-spring-2025/new-national-food-strategy
33 UK Gov, Policy paper: A UK government food strategy for England, DEFRA, 2025, https://www.gov.uk/government/publications/a-uk-government-food-strategy-for-england
34 Grylis, Bethany, UK government publishes 10 goals to transform Britain's food system, Food Manufacture, 2025, https://www.foodmanufacture.co.uk/Article/2025/07/15/uk-government-publishes-10-goals-to-transform-britains-food-system/
35 UK.Gov, Information paper food 2030, the advisory committee on the microbial safety of food, DEFRA, 2010, https://acmsf.food.gov.uk/sites/default/files/mnt/drupal_data/sources/files/multimedia/pdfs/committee/acm989food2030.pdf
36 Press Association, Sir Keir Stamer makes major cabinet reshuffle after Rayner quits over tax row, *insider media*, 2025, https://www.insidermedia.com/news/national/sir-keir-starmer-makes-major-cabinet-reshuffle-afterraynerquits-over-tax-row
37 Case, Philip, Angela Eagle replaces Zeichner as DEFRA farming minister, *Farmers Weekly*, 2025, https://www.fwi.co.uk/news/farm-policy/angela-eagle-replaces-zeichner-as-defra-farming-minister
38 Emma Reynold appointed new DEFRA secretary, NFU, 2025, https://www.nfuonline.com/updates-and-information/emma-reynolds-defra-secretary/
39 US.Gov, Report to the President on the America first trade policy: Executive summary, The White House, 2025, https://www.whitehouse.gov/fact-sheets/2025/04/report-to-the-president-on-the-america-first-trade-policy-executive-summary/
40 Atkinson, Robert, Trump the protectionist: Canada and Mexico are the first salvos, Information Technology Information Foundation, 2025, https://itif.org/publications/2025/02/02/trump-the-protectionist-canada-and-mexico-are-the-first-salvos/

41 Rivers, Dan, Donald Trump unveils 'liberation day' tariffs including 10% on British goods, ITVX, 2025, https://www.itv.com/news/2025-04-02/donald-trump-tariffs-liberation-day-announcement

42 US.Gov, President Trump's bold trade actions draws praise, The White House, 2025, https://www.whitehouse.gov/articles/2025/04/president-trumps-bold-trade-action-draws-praise/

43 News Wires, Possible pain from tariffs 'Worth The Price' Trump Says, France24, 2025, https://www.france24.com/en/live-news/20250202-trump-says-tariff-pain-will-be-worth-the-price

44 Goldmann, David, Trump's tariffs are inflicting serious economic damage and reigniting inflation, OECD says, CNN Business, 2025, https://edition.cnn.com/2025/03/17/economy/tariffs-oecd-forecast-economy-inflation/index.html

45 Mangan, Dan, Tariffs: Trump pauses tariffs for most countries, hits China harder with 125%, *CNBC*, 2025, https://www.cnbc.com/2025/04/09/trump-announces-90-day-tariff-pause-for-at-least-some-countries.html

1 The roots of our food crisis

So, what has led the UK to this food economy? How did we get to this moment? The primary policy focus that has driven us to this moment is a policy of delivering food security for the nation. What is food security? The term 'food security' was first formalised at the World Food Conference in 1974, which described it as the "availability at all times of adequate world food supplies of basic foodstuffs to sustain a steady expansion of food consumption and to offset fluctuations in production and prices".[1] Like governments across the world, the UK has worked to deliver food security for its citizens over the last century—with mixed success. During the war years, there was a push for production via agricultural intensification and farming supports to drive up yields. This was linked heavily to the war effort with the slogan 'Dig For Victory'. It saw yields prioritised over everything else and small pieces of marginal land and green space (famously including Parliament Square) converted into allotments. This was due to the naval blockades and vulnerabilities of shipping routes to U-Boat attacks. In response to the risk, the UK had to turbo charge their domestic food security since it could no longer rely on guaranteed imports to meet the demand for food. The 1939 Emergency Powers (Defence) Act enabled central control over farming, with the government fixing prices, guaranteeing markets, and heavily directing production.[2] The War Agricultural Executive Committee, known as the War Ags Programme, pushed to plough up pasture for arable crops and horticulture, distributed seeds and machinery, and encouraged farmers to embrace innovations. As a result, wheat production doubled during the war.

After the war, the decimated country still needed to be fed—there was a baby boom underway, and people wanted an end to the austerity and rationing. The farming intensification approach was seen as a successful path to follow. Consequently, the government passed the Agriculture Act in 1947, which continued the push for yields through the modernisation of farming.[3] The approach to increasing yields rested on the industrialisation of agriculture and the invention of agrochemicals. Synthetic fertiliser was used to make crops grow more efficiently. This was made possible through the wartime discovery of the Haber–Bosch method that allowed atmospheric nitrogen to be

DOI: 10.4324/9781003581871-2

converted into ammonia that plants could use to grow. Farmers were actively encouraged to apply fertilisers to their crops to drive up production. Further, there was a push on pesticides (an overarching term to include herbicides, fungicides, and insecticides). Chemicals like DDT and 2,4-D that had been used in the war as both weapons and to control the spread of disease within barracks were now applied to crops to reduce disease and pests. As a result, agricultural productivity was driven sky high. Between 1953 and 1973, in UK agriculture, agricultural production increased by approximately 50%.[4] Another key aspect of the industrialisation of UK agriculture after the Second World War was the improvement in plant breeding. Globally, this was pioneered by Norman Borlaug, who ultimately won the Nobel Peace Prize in 1970 for his work in delivering food security to Mexico and India.[5] He successfully bred a new variety of semi-dwarf wheat that was high-yielding and disease-resistant. In the UK, this meant the option of commercial varieties that grew fast and delivered uniform results. The push for the industrialisation of farming from 1945 onwards had a dramatic impact on small-scale farmers. Due to the focus on scale, there was a desire to push for bigger farms that could benefit from the economies of scale. Bigger fields to support bigger machinery to grow more crops in monoculture, which drives yields higher. Policies were enacted to support this transition, offering subsidies and tax incentives to large operations while making it harder for small farms to thrive, and small-scale family farms suffered. The UK has seen a significant decline in the number of small family farms over the last century. In 1939, there were nearly 500,000 farms in Britain, the majority of which were under 100 acres and operated by families. By 1990, this number had decreased to 319,000, and by 2001, it had further fallen to 168,000.[6]

This understanding of delivering food security through intensification continued. While post-war reconstruction laid the foundations, it was after the UK joined the European Economic Community (which later became the EU) in 1973 that agricultural intensification became firmly embedded in national policy. Under the CAP, farmers were rewarded directly for output, not environmental stewardship. This payment structure drove rapid expansion of monocultures, widespread use of synthetic fertilisers and pesticides, drainage of wetlands, and the concentration of livestock systems. This was a period where CAP's extensive focus on production and farming support led to overproduction and criticisms of 'butter mountains' and 'wine lakes', driven by the fact that farmers were guaranteed a fixed minimum price and therefore were incentivised to produce as much as possible. Mechanisation accelerated, and advisory services promoted high-yielding varieties and chemical inputs as the surest route to national food security. By the mid-1980s, UK agricultural productivity had risen by more than 30%, a reflection of how effectively policy redirected farming towards maximising production per hectare.[7] Admittedly, subsequent CAP reforms attempted to curb the environmental harms associated with early productivism. In recognition of the market-distorting problems of the CAP policies and their lack of a green agenda, reforms were made in the 1990s. Brussels moved to make fallow land pay, created environmental support payments, and ended guaranteed prices. They started to shift to focus on quality over quantity. In the UK, this resulted in the establishment of the Countryside Stewardship scheme, which was launched in 1991. This new scheme created tiers of support and capital grants to encourage farmers to make provision for nature within their land, such as hedgerows, wildflower meadows, and wetlands. However, the trajectory was already set, and the momentum towards intensification and commercialisation was building. The introduction of agri-environment schemes largely operated at the margins: the central policy logic continued to prioritise efficiency, yield, and the ability of UK agriculture to compete in global markets.

Industrial agriculture today

This focus on pushing agriculture towards productivity that has dominated the post-war era of agricultural policy continues today. Advocates point to the coming population boom and stress that a focus on production is essential if we are to feed each of those new mouths. The UN predicts that the world population will reach 8.5 billion by 2030, 9.7 billion by 2050 and 10.4 billion by 2100.[8] That is a lot of people who deserve to have ready access to food. Given this, many argue that crop innovation and improved breeding techniques are the best way to ensure that everyone can be fed. While environmental concerns have crept in, the same drive for food security delivered through industrialisation remains core. The language has changed with calls for 'sustainable intensification'—the aim of producing "more from less" by improving input efficiency, but the core vision for the future of farming remains the same. Post-Brexit reforms, particularly the ELM schemes, have been promoted as a shift away from output-based support towards a model of 'public money for public good', yet productivity remains a core objective. Government strategies emphasise technological innovation, including precision farming, automation, genetics, robotics, and data-driven input optimisation, as key pathways to securing the nation's food supply. Although framed as "sustainable", this orientation represents continuity with past approaches: intensification remains the primary mechanism through which the UK seeks to deliver food security, with ecological outcomes dependent on the hope that efficiency gains can be aligned with environmental protection.

Importing food security

Yet, the intensification of farming alone was not the only facet of the post-war drive for food security in the UK over the last century. Once the seas were cut once again safe for passage and free of threat, food imports resumed. The UK had long been a colonial power and had used its vast control of territories around the world to ensure a consistent supply of food for the British people. This move started with the repeal of the Corn Laws in 1846, which allowed for the flow of cheap grain from the rest of the world, and, as the Empire expanded, the UK relied on its colonies to feed it. Sugar from the West Indies, lamb from New Zealand, and grain for India. While the end of WWII saw the British Empire crumble, the mindset that food security could be imported did not end. The General Agreement on Tariffs and Trade in 1947 aimed to reduce tariffs and allow goods, including food, to flow freely, and ushered in an era of globalised free trade.[9] By 2023, the UK produced only 62% of the food it consumed.[10] Leading food expert, Professor Tim Lang, pithily noted this dynamic in his book, Feeding Britain: Our Food Problems and How to Fix Them, when he wrote "a post-imperial default policy still rules: that others can feed us". He continued,

> Britain's reflex is to let others feed it. This is wrong today, was wrong in the past, and is not just hugely risky in security terms, but betrays a malign assumption that we have right to be fed. We do not. This hangover from food imperialism is a bubble that needs to be burst.[11]

The risk of that bubble was exposed by Brexit. The UK's decision to leave the EU threw food imports into disarray. Prior to the Referendum, food imports from the EU to the UK were approximately 77%.[12] Following the UK's departure from the EU Customs Union on January 1, 2021, food and drink imports from the EU decreased to 22.5%.[13] Many Brexiteers expected that the difference could be met by imports from other nations outside of Europe. The UK was keen to establish new bilateral free trade agreements with some of the biggest food-exporting nations. The Tory Government were highly invested in securing these trade deals—in part, it was a way to justify the decision to Brexit. Prime Minister Boris Johnson expressed this commitment to international trade when he said, "we want to engage with the world again in a way that we haven't been able to do for 43 years".[14] The Australian Free Trade Agreement (FTA) was signed in 2021, and the New Zealand one was signed in 2022 with high hopes remaining for an FTA with Canada, the USA, and Brazil to gain access to their agricultural markets.[15,16] While he was Prime Minister, Rishi Sunak similarly worked to improve UK access to imports (food and otherwise) by securing the Windsor Framework that eased import restrictions with the EU.

Since Labour took office in 2024, the UK Government has continued to take a proactive approach to strengthening its global trade relationships, with a clear focus on improving access to diverse food markets. Labour achieved a key priority in May 2025 by signing a comprehensive trade agreement with

India, one of the world's largest food exporters, aiming to facilitate increased imports of key food commodities.[17] This move is intended to diversify the UK's food supply sources, reduce vulnerability to supply chain disruptions, and support consumer choice and affordability. Alongside these efforts, the Department for Business and Trade is also engaging with other emerging markets to broaden import options and strengthen resilience. Labour's trade strategy reflects a commitment to building a more interconnected and secure food system through expanded international cooperation and market access. A significant milestone was reached in 2025 with the successful negotiation of a trade pact with the USA. The initial iteration of this trade pact addressed the additional tariffs that President Trump has added to UK exports and reduced the burden.[18] They hope this pact will open a pathway to a bilateral trade deal between the UK and the US—something that had eluded the Tory Government. A trade deal remains a highly controversial proposition since there is strong concern within the UK about the standards and safety of US food production. Whether that fear is justified or not, opening up the UK food market to US imports will elicit a strong public reaction.

Box 1.1 Food security as a public good?

In economic terms, a public good is defined as a commodity or service that is both non-excludable and non-rivalrous—that is, individuals cannot easily be prevented from accessing it, and one person's use does not diminish its availability to others. Food itself does not meet these criteria, as it is inherently both excludable and rivalrous; thus, food production could not be directly subsidised under the proposed framework. However, the National Farmers Union contended that food security—as a condition beneficial to society as a whole—does meet these criteria, being both non-excludable and non-rivalrous. Despite this reasoning, the attempt to classify food security as a public good and to secure its inclusion within the ELM scheme was ultimately unsuccessful.

National food security vs. individual food insecurity

This approach of agricultural intensification and increased imports might have delivered national food security, but not individual food security. Over the years since the aforementioned 1974 definition of 'food security', there has been further discussion over how to understand food security. The early definition overlooked how food is distributed and whether people could actually obtain it. In response, by 1983, the Food and Agriculture Organization (FAO) refined the concept to include both physical and economic access to food.[19]

This shift acknowledged that availability alone was insufficient. If people lacked the means to purchase food, they would remain hungry. Food security was no longer just a matter of production, but also of ensuring individuals and households could meet their food needs. The concept was further expanded in 1986 when the World Bank released a pivotal report *Poverty and Hunger* and introduced a temporal dimension, distinguishing between chronic food insecurity, which is long term and rooted in poverty and low incomes, and transitory food insecurity, which arises from temporary shocks such as natural disasters, conflict, or economic crises.[20] This distinction helped policymakers better target long-term development and short-term humanitarian responses. At the same time, Amartya Sen's theory of entitlements (1981) revolutionised thinking about famine and food insecurity.[21] Sen argued that hunger was not always caused by a lack of food, but often by a failure of people's ability to access food. His work highlighted the importance of political, social, and economic systems in determining who eats and who does not. At the FAO World Food Summit in 1996, the contemporary understanding of food security was defined as, "when all people, at all times, have physical and economic access to sufficient safe and nutritious food that meets their dietary needs and food preferences for an active and healthy life".[22] The summit conceptualised four essential pillars of food security; *Availability* refers to the consistent supply of food through domestic production, imports, or food aid; *Access* involves having the necessary resources—whether income or infrastructure—to obtain appropriate foods; *Utilisation* focuses on the proper biological use of food, requiring a diet with sufficient energy and nutrients, clean water, and proper sanitation; and *Stability* ensures that people's access to food is secure over time and not disrupted by sudden shocks like conflict, economic crises, or natural disasters.

Individual food insecurity in the UK

By differentiating between food security for the nation and food security of the individual, questions of access, affordability, and inequality entered the discussion. While the UK Government focused on achieving national food security over the past century through agricultural intensification and imports, its action on individual food security was woefully lacking. This has led to shocking levels of hunger, with individual food security in the UK today being extremely poor. Shockingly, in 2024, one in five households cannot meet the basic food needs.[23] According to The Food Foundation, 9 million UK adults (17% of households) experienced moderate or severe food insecurity in June 2023, up from 7.3% in June 2021.[24] The primary drivers of this issue are low income and high living costs. Food banks have become a critical resource, with over 1.4 million emergency food parcels distributed between April and September 2024, including 508,000 for children.[25] To attempt to better tackle food insecurity in the UK, the Agriculture Act added a requirement to quantify food insecurity in the UK, legal requirement for the government to annually provide data on UK food security.[26] It requires Defra to prepare and lay before Parliament "a report containing an analysis on statistical data relating to food security in the United Kingdom" at least once every 3 years.[27] The first of these reports was published in December 2024.[28] The UK Food Security Report frames food security across six interconnected dimensions: availability, access, utilisation, stability, sustainability, and resilience. Recent shocks have tested each of these in turn, revealing both the system's capacity to recover and its underlying vulnerabilities. It made for sad reading. The most recent findings from their 2024 report showed the proportion of food-secure households declined from 92% in 2020 to 90% in 2023, exacerbated by inflation.[29]

Even for those individuals who are food secure, it does not mean that they have a healthy diet or sufficient nutrition. Food security is based on having sufficient calories. It does not consider the nutritional value of those calories. Having a healthy diet is a significant expense—an expense that many cannot afford. Healthy foods are nearly three times as expensive per calorie as less healthy foods.[30] Research by the University of Cambridge found that the poorest fifth of UK households would need to spend 47% of their disposable income on food to meet the cost of the Government-recommended healthy diet.[31] For those households with children, that figure increases to 70% of their disposable income to achieve a healthy diet.[32] Consequently, we face a crisis in public diets where it is cheaper and easier to buy chicken and chips than it is to buy fresh fruit and veg. Cheap food tends to be high in salt, fat, and sugar (HFSS) and/or ultra-processed (UPFs). These food groups are closely linked to dietary ill-health, obesity, and associated non-communicable diseases, such as diabetes.[33] These illnesses are rising dramatically in the UK.[34] The prevalence of diabetes in England has increased over the last 20 years, and the NHS is estimated to spend over £130 billion (175.9 billion dollars) annually on non-communicable diseases (NCDs).[35] The rise in NCDs is closely linked to poverty. Obesity rates are higher in more deprived areas. Specifically, 35.9% of adults in the most deprived areas were obese, compared to 20.5% in the least deprived areas.[36] Consequently, in 2023, the all-age cardiovascular disease (CVD) mortality rate was 51% higher in the most deprived decile than in the least deprived decile. The premature CVD mortality rate was twice as high in the most deprived decile compared to the least deprived decile.[37]

To address this, instead of questioning whether a food system that is dependent on intensification and imports is fit for purpose, Westminster has created bolt-on provisions that aim to redress hunger. These provisions do not challenge the model of intensification and imported food that has underpinned their food security policy over the last century. But rather, it acts as a band-aid for a failing food system. The Labour Government, under Tony Blair's leadership, launched the "Healthy Start" programme in England,

Wales, and Northern Ireland in 2006.[38] This programme supports pregnant women or families with children under 4 to purchase healthy foods, such as fresh, frozen, and tinned fruit, vegetables, and pulses—as well as milk and formula. The scheme continued under the Coalition and Tory Governments. The Healthy Start Scheme provides £4.25 each week of your pregnancy from the 10th week, £8.50 each week for children from birth to 1 year old and £4.25 each week for children between 1 and 4 years old. The scheme has had a positive impact. A study analysing data from 2021 found that households participating in the Healthy Start Scheme purchased an average of 0.9 more portions of fruit and vegetables per day compared to the baseline period in 2019.[39] This increase was statistically significant, suggesting that the scheme may help improve dietary intake among eligible families. However, despite its potential, Healthy Start only covers a slim proportion of the population, and many do not qualify. Sensible revisions to the scheme would help expand its impact. For example, Healthy Start values should keep pace with inflation, reflecting the cost-of-living crisis, and be expanded to all families on Universal Credit. Furthermore, some eligible families do not know that they can receive support. As of April 2024, the uptake rate was 62.4%, falling short of the government's target of 75%.[40] Auto-enrolment and better marketing would make a big difference to drive uptake.

Beyond Healthy Start, since 2014, England has provided universal free school meals to all children in Reception, Year 1 and Year 2, irrespective of household income. Children outside those years may still have qualified if their parents or carers met specific criteria tied to low income. However, education and welfare policy is devolved, so approaches vary across the UK. Scotland and Wales guarantee free school meals throughout primary school, while Northern Ireland offers free meals only to eligible children based on benefits. In June 2025, the UK Government announced an expansion of free school meals. From September 2026, all households on Universal Credit will automatically qualify for free school meals. This move is expected to add over half a million pupils to the scheme.[41] The expansion builds on evidence from the Universal Free School Meal pilot, which showed improved academic outcomes—particularly among low-income pupils.[42] Despite the significant benefits that universal free school meals might bring, the concept remains politically difficult and costly. A more feasible interim step would be expanding the offer to students in Years 3 and 4 who are in households on Universal Credit. Additionally, the Children's Wellbeing and Schools Bill, which is currently going through Parliament at the time of writing, includes a provision for free, universal primary school breakfast clubs (30 minutes before class) for all pupils at schools where 40% of the student body meet the Income Deprivation threshold.[43,44] This is designed to complement expanded lunch provision, deliver on Labour's 2024 Manifesto pledge, and help ensure that children never begin class hungry.[45]

Environmental damage

The side effects of the focus on agricultural intensification and the Green Revolution are not limited to health. It has also damaged the environment. The focus on increasing yields and driving production has had devastating impacts on our climate and our biodiversity. Modern industrial methods of agricultural production are key contributors to climate change and environmental degradation. There has been a catastrophic 73% decline in the average size of monitored wildlife populations in the last 50 years, according to the World Wildlife Fund *Living Planet Report 2024*.[46] It is estimated that agriculture is the primary driver of species decline for 86% of species that are currently threatened. Greenhouse gas (GHG) emissions from agriculture account for about 30% of global warming.[47] While agriculture is accelerating the rate of climate change, climate change is, in turn, wreaking havoc on farming and food production. Overall, crop yields are expected to decrease by up to 23% by the end of the century if global GHG emissions continue on their current track.[48] Hotter climates mean thirstier crops, but it also means less water. Approximately 76% of global croplands experience water scarcity for at least One month annually, and 42% face water scarcity for 5 months or more.[49] Increased and more extreme droughts may further destabilise crop production and dry out fertile topsoil, turning it to dust as it loses organic matter and carbon. By 2050, around 90% of the Earth's topsoil could be gone.[50] The push for more and more food production has also led to land use change as native forest is converted into farmland. This has exacerbated climate change, and has also driven up biodiversity loss. The delivery of food security rests on the industrialisation of agriculture and the use of agrochemicals. From the 1940s onwards, the use of pesticides increased dramatically, as availability increased and cost decreased. The trend continues. According to the FAO, global pesticide use in agriculture doubled from 1.85 million tonnes in 1990 to 3.70 million tonnes in 2022.[51] This increase reflects a 13% rise over the past decade and a 4% increase from 2021. The global trend of increasing pesticide use is expected to continue, especially in emerging agricultural markets where the drive to deliver food security remains top of the political agenda.[52] The use of these chemicals has decimated insect populations. Research indicates that nearly half of all insect species worldwide are experiencing rapid declines, with a third at risk of extinction.[53] In the UK, a shocking 80% of butterfly species have decreased since the 1970s.[54] Beyond pesticides, the use of synthetic nitrogen fertiliser also drives biodiversity loss as a result of leaching. Excess nitrogen can run off the soils into waterways, and the increase in the supply of nutrients can lead to excessive growth of phytoplankton and algal 'blooms'. When they die, it causes a reduction in oxygen levels, causing hypoxic zones. Without oxygen, other species cannot survive, and invertebrates and fish die off as the area of suitable habitat for food, growth, and reproduction becomes significantly reduced. The negative impacts can also move up the food chain to seabirds, mammals, and other marine animals. The number of dead zones has approximately doubled each decade since the 1960s, and they now affect more than 245,000 km^2 globally.[55] The Gulf of Mexico has seen the largest recorded dead zone of 8,776 square miles (roughly equivalent to the size of Wales) from fertiliser run-off washing out from the Mississippi.[56] Nitrates in water also damage human health and air quality.[57]

Food sovereignty

Given these problematic consequences of delivering 'national food security' through intensification and imports, it was inevitable that a counter movement would arise. This counter movement is known as food sovereignty. The movement first arose in Latin America in the 1980s and progressively expanded as peasants and family farmers formed alliances and began to organise on a large scale. In 1993, various groups cemented their collective efforts by forming *La Via Campesina*.[58] Food sovereignty rejects the FAO understanding of food as a commodity (and source of profit). Instead, the food sovereignty movement perceives food as, among other things, a human right and a source of nutrition and cultural value, inseparable from socio-ecological contexts. Instead, they push for food production where small-scale farmers produce nature-friendly food for their own communities and embrace the concept of social justice. Food sovereignty highlights the structural factors underlying global hunger and malnutrition, framing them as political, rather than technical issues.

Definitions of food sovereignty are multiple and constantly changing. The concept has grown beyond its original base among peasants in the Global South and has been adopted by the food justice movement in the Global North. It focuses on opposing the growing dominance of industrial agricultural practices, the increasing power of corporations in the global food system, and the convergence of diets towards processed foods. The most commonly accepted definition comes from the 2007 Nyéléni Forum:

> the right of peoples to healthy and culturally appropriate food produced through ecologically sound and sustainable methods, and their right to define their own food and agriculture systems. It puts the aspirations and needs of those who produce, distribute and consume food at the heart of food systems and policies, rather than the demands of markets and corporations.[59]

This definition acknowledges food sovereignty not only as a theoretical concept but also as a strategy for change. It focuses on 'people' rather than 'nations' and highlights the importance of environmental issues. That said, the movement tends to avoid defining specific visions of food sovereignty. Instead, it encourages members to embrace shared principles and adapt to their context.

How do we achieve this shift? If food sovereignty is the "why", then agroecology is the "how". Agroecology is the practice, while food sovereignty is the movement. They feed each other to build a more just and sustainable food future. So what is agroecology? Around the world today, the most commonly recognised definition of agroecology is that of the FAO, which has defined agroecology as "an integrated approach that simultaneously applies ecological and social concepts and principles to the design and management of food and agricultural systems", adding that "it seeks to optimise the interactions between plants, animals, humans and the environment while taking into consideration the social aspects that need to be addressed for a sustainable and fair food system".[60] The FAO definition is underpinned by ten principles: diversity, co-creation and sharing of knowledge, synergies, efficiency, recycling, resilience, human and social values, culture and food tradition, responsible governance, and circular economy.[61] Some of these elements are focused on the farming practices themselves, while others relate to society, values, and culture. The inherently holistic definition of agroecology and food sovereignty goes beyond farming to the wider society. It acknowledges that the social and environmental spheres are systemically linked and cannot be tackled in isolation. On-farm practices aren't enough to shift the food system. Wider social reform is needed to ensure that our food system can deliver for people and for the planet. This political core has been present within agroecology, and by extension, food sovereignty, since it was first conceptualised. At its conception, it presented itself as an alternative to the Green Revolution to combat the marginalisation of rural and Indigenous producers. Today, it envisions a more socially just food system at a global and local level and focuses on creating fairer conditions for agricultural and food-production workers.

Food sovereignty advocates the value of agroecological production to meet the challenge of a warming and increasingly unsettled climate. Agroecological farmers farms the land in harmony with nature, with the aspiration of helping the ecosystem heal. They use nutrient cycling (also known as rotational mixed farming) to restore soil health, rather than depleting soil nutrients through intensive cropping that relies on synthetic fertilisers. They argue that nature-friendly farming makes the land more capable of producing healthy food for longer since humans are not sucking all the goodness out of the soil. Furthermore, they argue that this approach can actually be carbon positive and help reduce GHGs by sequestering carbon into the soil, as opposed to merely trying to hit net zero, through carefully managed grazing. Globally, grasslands contain a third of all terrestrial carbon (90% in the soil).[62] These carbon sinks need to be protected. The FAO estimates that 33% of global soils are degraded and therefore have the potential to increase their carbon levels.[63] Holistic grazing (extensive systems on species-rich permanent pasture) can help to restore carbon into the soil, drawing it down from the atmosphere and locking it away. By allowing for natural nutrient cycling, cattle (and other ruminants) can help increase soil organic matter through their waste. The cattle feed on grasses and other plants that send their roots deep into the ground, furthering organic matter and allowing soil health to thrive. These sources of soil organic matter increase the soil carbon levels. This can be pushed even higher if you incorporate hedgerows and/or trees into the landscape. Through agroecological livestock management, agriculture can be a nature-based solution to climate change.

Box 1.2 Agroecology vs. regenerative agriculture

While regenerative agriculture and agroecology share aims of improving soil health, biodiversity, and resilience, they differ in scope and philosophy. Regenerative agriculture is primarily a management framework, focused on restoring ecological functions within existing market systems—often emphasising carbon sequestration, soil health, and input reduction. It is pragmatic, adaptable, and increasingly adopted by corporate actors. Agroecology, by contrast, is both a science and a social movement. It integrates ecological principles with questions of equity, power, and food sovereignty, seeking systemic transformation rather than incremental improvement. Where regenerative agriculture restores ecosystems, agroecology reimagines the entire food system as ecological, just, and democratic, including a focus on public health and fairness.

UK action on agroecology

The UK Government has started to wrestle with the conflict between national and individual food security, its shocking levels of hunger, and the growing movement towards food sovereignty and agroecology. A huge first step was the reference to 'Agroecology' in the Agriculture Act. Due to concerted lobbying by Sustain and the Land Workers Alliance, the Act included a direct reference to agroecology, specifically Clause 1(5) of the Act.[64] However, while food sovereignty groups welcomed its inclusion, many felt it should have gone further. For a start, Clause 1(5) of the Act states that the Secretary of State may provide financial assistance to achieve a "better understanding of agroecology". That is perhaps not such a strong mandate for pushing significant funding towards food sovereignty. Many challenged the weak language while the Bill went through both houses of Parliament. Lord Hain, for instance, argued that agroecology should be a central component of the Act, rather than a marginal note, to truly support sustainable farming practices.[65]

Westminster's acknowledgement of the failures of industrial agriculture intensification and imports to deliver food security continued with the (aforementioned) Dimbleby-led, independent National Food Strategy. While the term "food sovereignty" is not explicitly mentioned, the Strategy principles and recommendations align with several key aspects of food sovereignty, particularly in promoting local control over food systems, sustainable agricultural practices, and equitable access to food. It also explicitly acknowledges the need for agroecological farms that focus on lower intensity production systems to produce food in harmony with nature. Building on this momentum, the Defra-backed National Food Strategy, published in 2025, explicitly recognised the problems associated with a food system focused on yield, acknowledging that "it has become increasingly apparent that the food system is driving negative impacts on our health, environment, and resilience that undermine our food security".[66] They acknowledge the need to "broaden 'food security' beyond self-sufficiency" to a more holistic understanding of what the term means. In line with this, then Minister for Food Security, Daniel Zeichner, detailed the Government's aspiration to "deliver a whole system change that will help the amazing businesses that feed our nation to grow and thrive, which means more jobs and stronger local economies, while making it easier for families to eat and feel better". This shift in thinking hints at a new understanding of the need to reframe food security not merely as producing more, but as producing better, distributing more equitably, and regenerating ecological systems.

Notes

1 Menga, Marina, Food Security, Foresight, 2023, https://www.climateforesight.eu/seeds/food-security/#:~:text=In%20this%20context%2C%20food%20security,fluctuations%20in%20production%20and%20prices

2 GOV.UK, Emergency Powers Acts, UK Parliament, 1939, https://www.parliament.uk/about/living-heritage/transformingsociety/private-lives/yourcountry/collections/collections-second-world-war/second-world-war-legislation/emergency-powers-defence-act-c20-1940-/

3 GOV.UK Agricultural Act 1947, legislation.gov.uk, 1947, https://www.legislation.gov.uk/ukpga/Geo6/10-11/48/contents

4 Zayed, Yago, & Loft, Philip, Briefing paper: Agriculture: Historical statistics, House of Common Library, 2019, https://researchbriefings.files.parliament.uk/documents/SN03339/SN03339.pdf

5 Thomas, Peggy, Hero of the hungry: The life & work of Norman orlaug, Feeding Minds, 2022.

6 Vidal, John, Bleak future for small farms, *The Guardian*, 2001, https://www.theguardian.com/uk/2001/feb/28/ruralaffairs.footandmouth

7 Zayed, Yago, & Loft, Philip, Briefing paper: Agriculture: Historical statistics, House of Common Library, 2019, https://researchbriefings.files.parliament.uk/documents/SN03339/SN03339.pdf

8 UN, World population to reach 8 billion on 15 November 2022, Department of Economic and Social Affairs, 2022, https://www.un.org/en/desa/world-population-reach-8-billion-15-november-2022#:~:text=The%20latest%20projections%20by%20the,at%20that%20level%20until%202100
9 General Agreement on Tariffs and Trade, World Trade Organisation, 1947, https://www.wto.org/english/docs_e/legal_e/gatt47_e.htm
10 GOV.UK, United Kingdom Food Security Report 2024, DEFRA, 2024, https://www.gov.uk/government/statistics/united-kingdom-food-security-report-2024/united-kingdom-food-security-report-2024-theme-2-uk-food-supply-sources#sub-theme-1-food-sources
11 Lang, Tim, *Feeding Britain*. Penguin, 2021.
12 O'Carroll, Lisa, Brexit added nearly 6bn to UK food bills in the last two years, research finds, *The Guardian*, 2022, https://www.theguardian.com/politics/2022/dec/01/brexit-added-nearly-6bn-to-uk-food-bills-in-two-years-research-finds
13 GOV.UK, United Kingdom Food Security Report 2024, DEFRA, 2024, https://www.gov.uk/government/statistics/united-kingdom-food-security-report-2024/united-kingdom-food-security-report-2024-theme-2-uk-food-supply-sources#sub-theme-1-food-sources
14 BBC News, Boris Johnson: Brexit will enable UK to engage with the world, *BBC*, 2017, https://www.bbc.co.uk/news/uk-politics-40715005
15 GOV.UK, UK-Australia free trade agreement, Department for Business and Trade, 2021, https://www.gov.uk/government/collections/uk-australia-free-trade-agreement
16 GOV.UK, UK-New Zealand free trade agreement, Department for Business and Trade, 2021, https://www.gov.uk/government/collections/uk-new-zealand-free-trade-agreement
17 GOV.UK, UK concludes trade deal with India, Department for Business and Trade, 2025, https://www.gov.uk/government/news/uk-signs-trade-deal-with-india
18 The White House, Fact Sheet: US-UK Reach Historic Trade Deal, The White House, 2025, https://www.whitehouse.gov/fact-sheets/2025/05/fact-sheet-u-s-uk-reach-historic-trade-deal/
19 UN, Food Security Policy Brief, FAO, 2006, https://www.fao.org/fileadmin/templates/faoitaly/documents/pdf/pdf_Food_Securit y_Cocept_Note.pdf
20 The World Bank, A World Bank policy study: Poverty and hunger issues and options for food security in developing countries, World Bank, 1986, https://documents1.worldbank.org/curated/en/166331467990005748/pdf/multi-page.pdf
21 Sen, Amartya, *Poverty and famines: An essay on entitlement and deprivation*. Oxford Academic, 1983.
22 UN, Rome declaration on world food security, FAO, 1996, https://www.fao.org/4/w3613e/w3613e00.htm
23 Rackham, R., Rise in number of people facing hunger in the UK, *BBC*, 2025, https://www.bbc.com/news/articles/c5yqw5d00r0o
24 Food Foundation, Written evidence submission: EFRA committee: Fairness in the food supply chain, the Food Foundation, 2023, https://foodfoundation.org.uk/publication/efra-committee-fairness-food-supply-chain-written-evidence-submission
25 Trussell Trust, More than 1.42 million emergency food parcels distributed in the past six months, Trussell Trust, 2024, https://www.trussell.org.uk/news-and-research/news/more-than-1.42-million-emergency-food-parcels-distributed-in-past-six-months
26 Coe, Sarah, et al., Research Briefing: The Agriculture Act 2020, House of Commons Library, 2020, https://commonslibrary.parliament.uk/research-briefings/cbp-8702/
27 Coe, Sarah, et al., Research Briefing: The Agriculture Act 2020, House of Commons Library, 2020, https://commonslibrary.parliament.uk/research-briefings/cbp-8702/
28 GOV.UK, United Kingdom Food Security Report 2024, DEFRA, 2024, https://www.gov.uk/government/statistics/united-kingdom-food-security-report-2024/

united-kingdom-food-security-report-2024-theme-2-uk-food-supply-sources#sub-theme-1-food-sources

29 GOV.UK, United Kingdom Food Security Report 2024, DEFRA, 2024, https://www.gov.uk/government/statistics/united-kingdom-food-security-report-2024/united-kingdom-food-security-report-2024-theme-2-uk-food-supply-sources#sub-theme-1-food-sources

30 Food Foundation, New data: Government recommended diet costs poorest 5th of UK half their disposable income, the food foundation, 2023, https://foodfoundation.org.uk/press-release/new-data-government-recommended-diet-costs-poorest-5th-uk-half-their-disposable

31 Food Foundation, New data: Government recommended diet costs poorest 5th of UK half their disposable income, the food foundation, 2023, https://foodfoundation.org.uk/press-release/new-data-government-recommended-diet-costs-poorest-5th-uk-half-their-disposable

32 Food Foundation, New data: Government recommended diet costs poorest 5th of UK half their disposable income, The Food Foundation, 2023, https://foodfoundation.org.uk/press-release/new-data-government-recommended-diet-costs-poorest-5th-uk-half-their-disposable

33 Mutebi, N., Health impacts of ultra-processed foods. *Parliamentary Office of Science and Technology*, 2024, DOI:10.58248/PB59

34 Henney, A. E., et al., Ultra-processed food and non-communicable diseases in the United Kingdom: A narrative review and thematic synthesis of literature. *Obesity Reviews*, *25* (4), e13682, 2024, DOI:10.1111/obr.13682

35 Lloyd-Williams, Ffion, et al., The quest for effective and equitable policies to prevent non communicable diseases: Co-production lessons from stakeholders workshops, *PMC*, 2020, https://pmc.ncbi.nlm.nih.gov/articles/PMC9278532/

36 DEFRA, United Kingdom Food Security Report 2024, GOV.UK, 2024, https://www.gov.uk/government/statistics/united-kingdom-food-security-report-2024/united-kingdom-food-security-report-2024-theme-2-uk-food-supply-sources#sub-theme-1-food-sources

37 British Heart Foundation, Socioeconomic inequalities in heart and circulatory diseases in England: An analysis, 2025, https://www.bhf.org.uk/-/media/files/what-we-do/our-research/heart-statistics/health-inequalities/socioeconomic-inequalities-in-heart-and-circulatory-diseases-in-england.pdf?rev=d29adfff59124633a1127873016a367a&hash=FFE5E71E11B6A6939AB006FA41BB185F

38 Ferree, Tilda, et al., Briefing: Strengthening the health start scheme to protect children's health, *The Food Foundation*, 2025, https://foodfoundation.org.uk/sites/default/files/2025-05/Healthy%20Start%20Briefing%20April%202025%20FF%20and%20FSNT.pdf

39 Parnham, Jenny, et al., Is the healthy start scheme associated with increased food expenditure with low income families with young children in the United Kingdom?, *Research Gate*, 2021, https://www.researchgate.net/publication/357123696_Is_the_healthy_start_scheme_associated_with_increased_food_expenditure_in_low-income_families_with_young_children_in_the_United_Kingdom

40 Parnhan, Jennie, et al., Is the healthy start scheme associated with increase food expenditure in low-income families with young children in the United Kingdom? *BMC Public Health*, *21* (1), 2021, https://www.researchgate.net/publication/357123696_Is_the_healthy_start_scheme_associated_with_increased_food_expenditure_in_low-income_families_with_young_children_in_the_United_Kingdom

41 GOV.UK, Over half a million more children to get free school meals, Department of Education, 2025, https://www.gov.uk/government/news/over-half-a-million-more-children-to-get-free-school-meals

42 GOV.UK, Evaluation of the free school meals pilot, Department of Education, 2010, https://assets.publishing.service.gov.uk/media/5a7ae157e5274a34770e7c3f/DFE-RR227.pdf

43 GOV.UK, Children's well being and school bills, UK Parliament, 2025, https://bills.parliament.uk/bills/3909
44 GOV.UK, Exploring local income deprivation, Office for National Statistics, 2021, https://www.ons.gov.uk/visualisations/dvc1371/#/E07000223
45 Labour Party, Change: Labour party manifesto, Labour Party, 2024, https://labour.org.uk/wp-content/uploads/2024/06/Change-Labour-Party-Manifesto-2024-large-print.pdf
46 WWF, Living Planet Report 2024, WWF, 2024, https://livingplanet.panda.org/en-US/
47 Yang, Yi, et al., Climate change exacerbates the environmental impact of agriculture. *Science, 385* (6713), 2024, https://www.science.org/doi/10.1126/science.adn3747
48 Rezaei, Ehsan Eyshi, et al., Climate change impact on crop yields. *Nature Reviews Earth and Environment, 4*, 2023, https://www.nature.com/articles/s43017-023-00491-0#citeas
49 Rosa, Lorenzo, et al., Global agricultural economic water scarcity. *Science Advances, 6* (18), 2020, https://www.science.org/doi/full/10.1126/sciadv.aaz6031?
50 Heggie, Jon, Why soil matters (and what we can do to save it), *National Geographic*, 2025, https://www.nationalgeographic.com/environment/article/paid-content-why-soil-matters-and-what-we-can-do-to-save-it
51 UN, Pesticide use and trade 1990–2022, FAO, 2024, https://www.fao.org/statistics/highlights-archive/highlights-detail/pesticides-use-and-trade-1990–2022/en?
52 Shattuck, A., et al., Global pesticide use and trade database (GloPUT): New estimates show pesticide use trends in low-income countries substantially underestimated. *Global Environmental Change, 81*, 102693. 2023, DOI:10.1016/j.gloenvcha.2023.102693
53 UN, Insect declines are a stark warning to humanity, *UNEP*, 2019, https://www.unep.org/news-and-stories/story/insect-declines-are-stark-warning-humanity?
54 Gilburn, A. S., et al., Are neonicotinoid insecticides driving declines of widespread butterflies? *PeerJ, 3*, e1402, DOI:10.7717/peerj.1402
55 UN, Issue brief – Ocean Hypoxia Dead Zones, UNDP, 2015, https://www.undp.org/publications/issue-brief-ocean-hypoxia-dead-zones
56 National Oceanic Atmospheric Administration, Gulf of Mexico 'dead zone' is the largest ever measured, 2017, https://www.noaa.gov/media-release/gulf-of-mexico-dead-zone-is-largest-ever-measured
57 Ward, Mary, et al., Drinking water nitrate and human health: An updated review. *International Journal of Environmental Research and Public Health, 15* (7), 2018, https://www.mdpi.com/1660-4601/15/7/1557
58 La Via Campesina, What is food sovereignty?, https://viacampesina.org/en/what-is-food-sovereignty/
59 Nyeleni, Declaration of Nyeleni, 2007, https://nyeleni.org/IMG/pdf/DeclNyeleni-en.pdf
60 UN, The 10 elements of agroecology, FAO, 2018, https://openknowledge.fao.org/server/api/core/bitstreams/3d7778b3-8fba-4a32-8d13-f21dd5ef31cf/content
61 UN, The 10 elements of agroecology, FAO, 2018, https://openknowledge.fao.org/server/api/core/bitstreams/3d7778b3-8fba-4a32-8d13-f21dd5ef31cf/content
62 Bai, Youngfei, & Francesca, Cotrufo, Grassland soil carbon sequestration: Current understanding, challenges, and solution. *Science, 377* (6606), 2022, https://www.science.org/doi/10.1126/science.abo2380
63 UN, Global Symposium on Soil Erosion: Key Messages, FAO, 2019, https://www.fao.org/about/meetings/soil-erosion-symposium/en/
64 Legislation.gov.uk, Agriculture Act 2020, UK Parliament, 2020, https://www.legislation.gov.uk/ukpga/2020/21/section/1

65 Parliament.UK, Agriculture Bill Vol804: Debated on Thursday 9 July 2020, Hansard, 2020, https://hansard.parliament.uk/Lords/2020-07-09/debates/0F2A8115-10FC-4BB6-A4EE-D3AC12DFC9B8/AgricultureBill
66 GOV.UK, A UK government food strategy for England, considering the wider UK food system, DEFRA, 2025, https://www.gov.uk/government/publications/a-uk-government-food-strategy-for-england/a-uk-government-food-strategy-for-england-considering-the-wider-uk-food-system#ministerial-foreword

2 Food resilience—a new focus

So, as discussed in the previous chapter, the UK Government's approach to the food system has been focused on driving food security through intensification and international trade over the last century, with an increasing recognition in more recent years, even at the highest levels. However, that is not working and a shift towards an agroecological approach is needed. Adopting an integrated approach that balances food security and food sovereignty is particularly essential since we are facing substantial challenges to our food system currently. These include climate change, rising geopolitical risk with trade wars and armed conflicts, and a public diet crisis. We need to build a food system that embraces the goal of food security while also recognising the need for improved food sovereignty. We need a system that aspires to both—embracing where they overlap and where they differ. To achieve this balance, we need a food policy that actively works to advance food security AND food sovereignty. Yet, there is still something missing—a critical gap between food security and food sovereignty. We need food systems that are better able to endure crises. It isn't sufficient to sit back and wait to see whether our food supply system can rise to the challenges that will inevitably come. Ensuring consistent access to safe and healthy food is essential, and it is paramount that we work proactively to ensure that our food system can withstand future crises. Resilience is the cornerstone of an enduring and robust food system. Our globalised food system is so precisely engineered to maximise profit that its weaknesses often go unnoticed—until they don't. It only takes one disruption for the entire chain to falter. These vulnerabilities can remain hidden until stress is applied. As we've discussed, countries like the UK have grown increasingly dependent on imports, with food supply chains stretching across continents. While imports can enhance food security, they also expose us to a web of global risks: rising geopolitical tensions, the weaponisation of food, population growth, and a slowdown in productivity gains. Together, these forces are putting future global food supply under growing pressure. As global supply becomes more unstable and climate change intensifies, domestic resilience becomes not just a preference but a necessity. Relying on production

DOI: 10.4324/9781003581871-3

that takes place thousands of miles away creates inherent fault lines in the system. Sometimes those cracks lead to small tremors—noticeable but manageable. A poor harvest might push up the price of tomatoes, prompting some to take note. But other times, the shake is unavoidable. COVID-19 was one such jolt. Around the world, food systems shuddered. And yet, as alarming as the pandemic was, it was only a warning tremor. The Big One—the event that truly tests or breaks the system—may still lie ahead. And we'd better be ready.

Yet, it is incredibly difficult to define something as ambiguous and intangible as resilience. We need resilience built into our food system, but what does that mean? Defining resilience in the food and farming sector is challenging due to its multidimensional nature. It spans ecological, social, economic, and institutional systems, making a single, universal definition difficult. Different stakeholders—farmers, businesses, and policymakers—prioritise different outcomes, such as climate adaptation, market stability, or food security. Resilience also operates across scales, from individual farms to global supply chains, and must account for both sudden shocks and slow-onset stresses. Additionally, resilience is dynamic; what works in one context may not apply elsewhere. As a result, definitions often vary, making measurement, comparison, and implementation complex.

The fault lines

So where might this big quake come from? What is causing the weaknesses in our food system? COVID-19 demonstrated just how fragile the UK food retail with shortages of staples such as pasta and rice and with chicken breast flying off the shelves. Consumers stared at shelves devoid of fruit or veg when demand for groceries soared in March 2020. People started to panic about the threat of a lockdown, amounting to the biggest month of grocery sales ever recorded in the country.[1] Many of the shortages seen across supermarkets at the early stages of the pandemic were due to stockpiling with people searching for a greater sense of security in a time of crisis—A very human response at a time when we all felt scared of what the future held. However, the speed at which basic essentials were depleted highlighted the reality of lean supply chains and just-in-time inventories that supermarkets have adopted the world over.[2] In the UK, the supermarket's reliance on this kind of supply chain means that the country only maintains 5–10 days' worth of groceries, and often less in the case of fresh produce.[3] In response to the growing crisis caused by the initial lockdown, the Defra launched the Food Resilience Industry Forum (FRIF) to support the logistical and technical operations of food supply across the UK food supply chain and relaxed competition regulations so supermarkets could better collaborate.[4] Parliamentary reporting acknowledged that the effect of consumers stockpiling certain goods and the slow reaction of retailers to ration them exposed the limitations of cost-efficient and streamlined supply chains to be agile and adapt to unforeseen shocks. This suggests that changes may be needed to make the supply chain more resilient. Specific problems arose from the closure of parts of the catering sector and the lack of agility in redistributing supplies from this sector to retail outlets or the food donation/charity sector. This was due to challenges in packaging availability, logistics and labelling requirements; leading to an increase in food loss and concluded that "the longer-term food system issues highlighted as a result of the pandemic will have to be addressed by considering how to build resilience to possible future shocks"[5].

One of the main risks to food stability is climate change and its unavoidable impact on food production. Farmers have been increasingly worried about the potential impact on their businesses, and many have been working to adapt their farming systems to be more resilient. However, much of this is obscured to the consumer, who doesn't notice much of a difference at the grocery store. The majority of consumers are blithely unaware of how climate change is impacting their food supply. Sure, some prices have started to rise, but most put that down to a poor economy. The true fragility of the food system facing a climate crisis is still obscured. It isn't until the system goes into crisis that people face that reality. For example, when Hurricane Sandy hit New York City in 2011, the city faced significant food shortages.[6] Some of the world's most affluent consumers had to directly confront the precarious position that extreme weather events can place on food supply. Similarly, 2023 saw the UK supermarkets face shortages due to inclement weather impacting food imports.[7] When asked to comment in Parliament, the Defra Secretary Thérèse Coffey said Asda, Morrisons, Aldi and Tesco were applying item limits to a small number of fruits and vegetables in response to issues with supply from Spain and north Africa caused predominantly by seasonal weather hampering production and harvest during December and January. The nature of horticulture and the effect on production of short-term events such as weather can create some volatility, and any growing forecast is subject to short-term alterations. We know that Ireland and other parts of Europe are facing very similar supply issues.[8]

While the environmental crises confronting us are profound, they are not the only challenges destabilising our food system. The past decade has been marked by growing geopolitical turbulence, with domestic politics and international tensions increasingly shaping global food flows. For the UK, Brexit created years of uncertainty and disruption, exposing the vulnerabilities of our reliance on cross-border trade and sparking ongoing debates over the status of Northern Ireland. Yet Britain is not alone: the surge of populism worldwide has fuelled protectionist policies and reshaped trade dynamics. Most notably, the introduction of tariffs by the Trump administration in 2025 has dramatically impacted global trade and rattled supply chains.[9] More destabilising still are the conflicts that have weaponised food on the global stage. Russia's full-scale invasion of Ukraine devastated agricultural production in one of the world's great breadbaskets, with blockades of Black Sea ports and deliberate targeting of farmland causing cascading shortages of grain and oilseeds.[10,11] Russia has openly admitted to using food as a geopolitical tool. In April 2022, former President Dmitry Medvedev referred to food as Russia's "silent weapon", and Ukrainian Minister of Agriculture, Vitaly Koval, described the attacks as "a big threat to us", saying they are intended to cut Ukraine's exportable supply while pushing up global prices to benefit Russian farmers.[12,13] Meanwhile, the conflicts in Gaza and Iran has underscored how access to food can be leveraged in war. Israel's military strategy has included tight controls on the flow of goods, in particular food, into Gaza, citing the need to weaken Hamas by cutting off its resources.[14] Since the conflict began, Israel has restricted or periodically halted the entry of food, fuel, and humanitarian supplies through key border crossings, most notably Rafah and Kerem Shalom, though humanitarian organisations have urged that "the collective punishment" of civilians violates international law.[15,16] Israeli officials have described these measures as part of a broader effort to stop Hamas from diverting aid from civilians to meet their own needs.[17] Still, the blockade has also impacted the civilian population, the majority of whom are already dependent on aid.[18] Restrictions on humanitarian aid have left Gaza on the brink of famine.[19] In today's interconnected world, such as the closure of the Straits of Hormuz, ripple far beyond their borders, disrupting supply chains, fuelling price spikes, and pushing millions more into hunger.

Defining resilience

The FAO defines resilience as

> The ability to prevent disasters and crises as well as to anticipate, absorb, accommodate or recover from them in a timely, efficient and sustainable manner. This includes protecting, restoring and improving livelihood systems in the face of threats that impact agriculture, nutrition, food security and food safety.[20]

This definition emphasises not just recovery, but proactive preparedness and sustainability, placing resilience as a long-term systems attribute rather than a short-term response mechanism. The UN, FAO's parent organisation, defines resilient societies as

> The ability of individuals, households, communities, cities, institutions, systems and societies to prevent, anticipate, absorb, adapt, and transform

> positively, efficiently and effectively when faced with a wide range of risks, while maintaining an acceptable level of functioning and without compromising long-term prospects.[21]

Here, the inclusion of "transform positively" suggests that resilience is not about returning to the previous state, but about improving and evolving in the face of adversity. An interesting additional distinction on the understanding of food system resilience within the UN sphere comes from the United Nations Convention to Combat Desertification, which acknowledges that "Resilient food systems are grounded in farmers who care for the land, work the fields, and take animals to pasture…".[22]

For food businesses, resilience is intertwined with corporate responsibility and sustainability. Nestlé is widely recognised as one of the world's largest food producers—in 2023, the company reported total sales of CHF 93.0 billion (approximately $105.5 billion)—and so it's worth considering how they define resilience as a wider indicator of agribusinesses' approach.[23] Nestlé's Creating Shared Value (*CSV*) strategy emphasises building "a resilient food system that replenishes the soil, supports livelihoods … and enhances the health of the environment".[24] Their focus reflects a pragmatic, supply-chain-centred definition of resilience—one that ensures business continuity while also supporting farmer adaptation and ecological regeneration. This positions resilience as an outcome of regenerative farming, long-term supply stability, and healthier ecosystems. It does little to acknowledge resilience on the consumption side. There is no acknowledgement of the role in resilience within access or supply chains—let alone in public health. Another representative voice for business worth consulting is the British Standards Institution (BSI). BSI is the leading UK voice on business, working to provide technical standards and guidelines that help organisations ensure quality, safety, and efficiency across various industries, including food. They focus on promoting consistency and innovation while reducing risks, supporting organisations in becoming more resilient and sustainable. In the food sector, BSI plays a key role by developing frameworks that enhance food safety and supply chain integrity. They take a purely business focus in their definition of resilience and frame organisational resilience as "the ability of an organization to quickly adapt to disruptions while maintaining continuous business operations".[25] They emphasise embedding resilience through standards, culture, and peer learning across the whole supply chain. This definition reflects a corporate management perspective, where resilience is embedded in risk management, standards compliance, and adaptive capacity within supply chains.

These differing definitions show us that resilience is not a fixed concept—it is shaped by context, scale, and values. For some, it is about bouncing back to a previous state after a disturbance; for others, it's about moving forward—transforming and evolving to better withstand future challenges.

Some definitions emphasise the resilience of the individual farmer or farm, focusing on local knowledge, adaptive capacity, and livelihood security. Others consider the entire global supply chain, addressing systemic risks such as market volatility, climate change, and geopolitical disruptions. The challenge lies in reconciling these diverse perspectives to create policies and practices that are inclusive, flexible, and forward-looking. Ultimately, the diversity of definitions signals that resilience must be both multidimensional and adaptable—just like the complex, interconnected food systems it aims to strengthen.

Agroecology and resilience

As mentioned earlier, one of the ten elements of agroecology, as defined by the FAO, is resilience. They define resilience as "the ability of agroecosystems, communities, and food systems to absorb shocks and stresses—whether environmental, economic, or social—and continue to function and evolve. It means bouncing back from disruptions while adapting and transforming over time".[26] However, the way in which they view 'resilience' is that it is a factor within agroecology. The FAO talk of how, "diversified agroecological systems are more resilient – they have a greater capacity to recover from disturbances including extreme weather events such as drought, floods or hurricanes, and to resist pest and disease attack". They continue by recognising that "agroecological approaches can equally enhance socio-economic resilience. Through diversification and integration, producers reduce their vulnerability should a single crop, livestock species or other commodity fail. By reducing dependence on external inputs, agroecology can reduce producers' vulnerability to economic risk".[27] This is all true, and it is worth recognising how agroecology can help to deliver resilience, but resilience is too important to be seen as a facet within something else—especially when our food system is as vulnerable as it currently is. The FAO presents 'resilience' as a benefit that flows from adopting agroecology—but this fails to fully recognise and acknowledge the value of resilience in its own right.

Has the UK considered resilience in food?

In 2006, the Defra commissioned a report titled *Resilience in the Food Chain* to assess the ability of the UK food system to withstand and recover from disruptions.[28] The study examined the extent to which businesses in the food and drink industry had adopted Business Continuity Management (BCM) practices. It found that while larger companies often had robust contingency plans, many small- and medium-sized enterprises were underprepared for crises. The report highlighted the vulnerability created by complex global supply chains and widespread reliance on just-in-time delivery systems, which reduced stockpiles and increased exposure to shocks such as fuel shortages, labour strikes, or transport disruptions. Communication gaps between government and industry were also noted, particularly in times of emergency. The report recommended expanding the use of BCM across the sector, strengthening communication networks, and re-evaluating efficiency-driven practices that compromise resilience. It served as an early warning of systemic fragilities and urged a more coordinated approach to food system security.

In hindsight, its findings proved prescient during events like Brexit and the COVID-19 pandemic, when the food supply chain faced significant stress, underscoring the ongoing need for resilient infrastructure and strategic preparedness in the UK food sector. Combined with this, the UK has an official *Government Resilience Framework*, most recently updated in 2022, as part of an evolving approach to national risk.[29] However, this framework gives minimal attention to food or the role of the public in supporting civil food resilience. Similarly, the 2023 National Risk Register identifies only one direct food-related threat: food supply contamination.[30] While this is a legitimate concern, it represents just one of many potential food system shocks. The limited scope of these assessments contrasts sharply with the concerns of food sector experts and industry leaders, who anticipate a broader range of risks. This suggests a significant gap in government thinking around food security and resilience.

This has culminated in the UK Government explicitly recognising the need for greater resilience to deliver food stability in their own 2025 National Food Strategy, which was introduced in Chapter 1. The strategy identifies four systemic challenges. First is public health, with obesity and diet-related disease placing immense costs on individuals and the NHS. Second is the environmental impact of agriculture and food supply, including biodiversity loss, soil degradation, and greenhouse gas emissions. Third are market failures and negative feedback loops, such as the "junk food cycle", where unhealthy, ultra-processed foods remain cheap, profitable, and heavily marketed. Fourth, and of primary concern for this book, is resilience and food security. The Strategy says that, as rising food prices, heavy reliance on imports, and exposure to global shocks threaten stability, we need a food system that "supports stability, enables economic growth and protects the most vulnerable" and, that by embedding resilience into supply chains, we can ensure not only food security but also market stability for producers and consumers in times of crisis.[31] The document acknowledges the 'resilience gap' given the "increasing risk from the intensifying impacts of climate change and interactions with heightened geopolitical tensions" and aims to shift policy towards a food economy that "is more resilient to short term shocks and better adapted to the long term challenges of climate change and biodiversity loss".[32] More importantly from the perspective of future policy development, the strategy specifically acknowledged Government's role in bridging that resilience gap, saying, "while there are incentives for businesses to build their own resilience, and many are, the incentives to take coordinated action to manage risks and strengthen resilience across the system, particularly for the poorest in society, are weaker".[33] It is notable that the language in the Strategy frames 'resilience' as geopolitical and/or economic. For example, it reads, "investment will deliver improved health, environment and resilience outcomes".[34] It fails to see the interconnected nature of these three. We need environmental and public health resilience as much as we need resilience within our supply chains. In fact, so notably does it separate resilience of the supply chain from the other aspects of the food system that it is its own target, along with "a thriving UK food sector", "a healthier population", and "better environmental outcomes".[35] While a dedicated focus on resilience is welcome, the siloed thinking on what constitutes a resilient food system and the exclusive focus on resilient supply chains are concerning. This siloed thinking has led to the mistaken belief that this is an easy challenge to fix. The Strategy claims, "Some issues, for example strengthening the resilience of some supply chains, may be more straightforward to resolve than others".[36] This is mistaken—something as complex and intricate as 'resilience' is not easy to fix, not least because it is such a complex and intricate concept to understand. If one struggles to define resilience, how can one hope to write it into policy?

National Preparedness Commission

The need to focus on resilience and the challenge of defining it has caught the attention of one of the greatest thought leaders in the food and farming sector: Professor Tim Lang, whose wise words were referenced previously. As a result of his standing in the sector, he was asked by the National Preparedness Commission (NPC) to dive into the current risks facing the UK food system. The NPC is an independent and non-political body whose fundamental objective is to promote policies and actions to help the UK be significantly better prepared to avoid, mitigate, respond to, and recover from major shocks, threats, and challenges. The goal was to consider policy change within the food and farming sector that will enable the UK to prevent, withstand, recover and 'build back better' from major disruptions to our food supply. Lang defined 'civil food resilience' as

> the capacity of people in their daily lives to be more aware of risks to food, more skilled in reducing unnecessary risks, and more prepared to act with others to ensure all society is well fed in and after crises.[37] This cannot be left to the people to activate in an ad hoc way or on their own. It requires a process of learning, capacity-building and preparation.[38] To be, as government wishes, a response of the 'whole of society' requires infrastructure, guidance and support.[39]

Drawing on interviews with over 70 experts from government, industry, academia, community groups, and Local Resilience Forums, analysis from ten other nations, and lessons from recent global shocks—COVID-19 and the Ukraine war—Professor Lang concluded that current resilience is woefully inadequate and calls for the UK to overhaul how it approaches food system resilience. The strategy warns against overreliance on centralised, just-in-time supermarket supply chains. It advocates for more diverse, decentralised systems—including local producers, smaller retailers, and community food initiatives—to build flexibility and redundancy. It also stresses the need for reviewing the UK's emergency logistics: stockpiling key staples, developing fair rationing protocols, and ensuring rapid food distribution in times of crisis. At its heart is a push for legislation that formally recognises food as critical national infrastructure. This would ensure coordinated leadership, dedicated resources, and strategic long-term planning at the highest levels of government. Local Resilience Forums would be required to include food in their emergency scenarios, mapping community resources and preparing region-specific contingency plans. One key recommendation is for a National Food Security Council that would unify responses across government, industry, and civil society, ensuring a joined-up approach to resilience.

His report outlines seven strategic steps to transition towards a "just-in-case" mindset, centred on seven core themes:

1. *Legislation and National Leadership:*
   The UK currently lacks a legally binding framework that treats food security as a core element of national resilience. To close this critical gap, the NPC recommends enshrining food resilience into law through a dedicated Food Security and Resilience Act. This would elevate food to the status of critical national infrastructure—on par with energy or communications—ensuring cross-government accountability, clear roles, and strategic oversight. Central to this proposal is the appointment of a named minister or coordinating body with the authority and responsibility to lead on civil food resilience. Such legislation would not only mandate coordination between departments but also facilitate long-term investment in preparedness, from emergency food stockpiles to early warning systems. By placing food at the heart of the UK's risk planning architecture, this recommendation aims to move the country away from reactive crisis response and towards a sustained, whole-society approach to protecting access to food during disruption.

2. *Local Resilience Integration:*
   To strengthen food system preparedness at the community level, the report calls for food resilience to be explicitly embedded in the UK's local emergency planning structures. Local Resilience Forums (LRFs), which already coordinate civil protection efforts, should be mandated to assess risks to local food access, map community food assets, and develop area-specific response plans. This approach acknowledges that food shocks—whether caused by weather events, economic volatility, or supply chain disruptions—often have uneven regional impacts. By empowering

LRFs with the tools, data, and authority to plan for local contingencies, the UK can build a more responsive and adaptive food system. This includes scenario planning, engagement with local producers, and partnerships with charities and food hubs. Strengthening local capacity ensures that national efforts are matched by grassroots readiness, making communities better equipped to maintain food access during times of crisis.

3 *Diversified Supply Chains:*
The UK's current food distribution system relies heavily on a centralised, supermarket-dominated model that prioritises efficiency over resilience. This approach leaves the nation vulnerable to systemic shocks—be it fuel shortages, border delays, or cyberattacks on logistics providers. The NPC proposes a shift towards a more diverse and regionally distributed supply chain that incorporates independent retailers, food co-ops, local producers, and community growing schemes. By decentralising food infrastructure, the UK can create redundancy and flexibility, ensuring that food can still move even if parts of the system fail. This diversification also has co-benefits: it supports rural economies, reduces environmental footprints, and increases consumer choice. More resilient supply chains are not just about more storage—they're about designing a food system that can bend without breaking. Investing in distributed networks will help ensure continuity of food access in emergencies while promoting a healthier, more robust food economy.

4 *Stockpiling, Rationing, and Logistics Planning:*
The UK has no clear public policy on food stockpiling or rationing for civil emergencies—unlike many other countries that maintain strategic reserves of essential goods. The NPC recommends that the government undertake a full review of food contingency measures, including the feasibility of national or regional food stockpiles, mechanisms for emergency distribution, and protocols for temporary rationing. This planning must include not only physical assets, such as warehouses and transport, but also digital tools for logistics coordination and equitable access. In moments of extreme disruption—pandemics, geopolitical shocks, or climate events—this infrastructure becomes critical. Without pre-planned logistics and supply strategies, the UK risks food price spikes, panic buying, and disproportionate impacts on vulnerable populations. The goal is not to replicate Cold War-era bunkers, but to develop a smart, flexible system that can pivot quickly to protect public access to food during crises, especially in areas with fewer local resources.

5 *Public Engagement and Civil Society Mobilisation:*
Food resilience is not just a matter for government and industry—it requires an informed and engaged public. The NPC emphasises the importance of civil society's role in preparedness, advocating for a broader public understanding of food system vulnerabilities and the actions individuals

and communities can take to mitigate them. This includes public education campaigns, community growing initiatives, and the integration of local food groups into emergency planning. Trusted NGOs and grassroots organisations often have the networks and local knowledge to respond quickly in a crisis, particularly for vulnerable or hard-to-reach populations. By empowering these actors and investing in their capacity, the UK can build social infrastructure that complements physical supply chains. Mobilising civil society also helps foster a sense of shared responsibility and readiness, moving the conversation away from top-down interventions towards collective resilience and solidarity in the face of disruptions.

6 *Cross-Sector Governance:*
Strengthening food resilience requires joined-up leadership across government, business, science, and civil society. To this end, the NPC proposes the creation of a National Food Security Council, bringing together representatives from all relevant sectors to coordinate policy, share intelligence, and plan collectively. This body would serve as a hub for scenario planning, crisis simulation, and strategic oversight, ensuring a consistent and unified response to food-related risks. The current governance landscape is fragmented, with no single entity accountable for the resilience of the entire food system. By contrast, a council structure would break down silos and encourage systems thinking, enabling decision-makers to understand the cascading effects of shocks—from agriculture through to public health. Such an approach reflects best practices seen in other national risk frameworks and would dramatically improve the UK's ability to anticipate and respond to complex food crises with speed and coherence.

7 *Data, Insight, and Learning:*
Effective resilience planning depends on accurate, real-time data and a culture of continuous learning. The NPC recommends the development of robust food resilience metrics and live monitoring systems to track vulnerabilities in supply, price fluctuations, regional disparities, and the impact of shocks. This would allow for more agile responses and better-informed policymaking. In addition, post-crisis evaluations should become a routine part of emergency management, feeding lessons learned back into future planning. Too often, the UK has responded to food-related disruptions—such as during COVID-19—without systematically capturing what worked, what didn't, and why. Creating institutional mechanisms for reflection and improvement would not only help avoid repeated failures but also build public trust. Transparent data sharing between sectors, accessible dashboards, and a national risk observatory for food would support this goal, laying the foundation for a smarter, adaptive, and forward-looking food security strategy.

Prof Lang's report for the National Preparedness Commission issues a stark warning: unless the UK rapidly adopts its proposed food resilience measures, it risks repeating past failures—such as empty supermarket shelves, sudden price spikes, and significant impacts on public well-being.[40] The call is for a shift from reactive crisis response to proactive planning, underpinned by legal mandates, local preparedness, and national leadership that treats food security as essential to national resilience. However, the NPC's focus remains largely on supply chain vulnerabilities—specifically, the risks inherent in just-in-time logistics and the concentrated power of major supermarkets. While this is a critical area for reform, it represents only a partial view of the challenge. To achieve meaningful and lasting food system resilience, our focus must move beyond logistics. Resilience must encompass environmental sustainability, land use, nutritional health, labour conditions, affordability, and equity in access. The system's fragilities are not just operational—they are social, ecological, and economic. A future-fit food strategy must recognise resilience as a multidimensional concept that addresses deeper structural risks, from climate impacts on agriculture to inequality in food access. This broader view will help ensure the UK's food system can not only withstand shocks but also adapt, recover, and serve all communities fairly and sustainably.

International insight

When Professor Lang reviewed the UK's approach to food resilience in comparison with other nations, he found a common pattern: most governments treat food resilience primarily as a disaster management issue. Their policies largely focus on preparing for and responding to acute crises, operating under the assumption that serious disruptions will be relatively short-term emergencies. This emphasis on immediate disaster preparedness shapes strategies, centred on emergency stockpiling, supply chain protection, and rapid response measures. While this approach strengthens readiness for sudden shocks, it often overlooks the importance of longer-term resilience-building, such as enhancing local food systems, supporting community-level adaptability, and addressing systemic vulnerabilities. By focusing predominantly on crisis response, many national plans miss opportunities to develop more robust, flexible food systems capable of withstanding a wider range of challenges over time. Key examples include:

- *Switzerland*
  Switzerland has prioritised preparedness for centuries, emphasising stockpiling and self-reliance during peacetime. As a landlocked country heavily dependent on imports—only about half of its food is domestically produced—Switzerland faces challenges due to reliance on imported seeds, fuel, fertilisers, and pesticides. The Swiss Federal Office for National Economic Supply (FONES), in partnership with the private sector, manages

compulsory stockpiles mandated by the Economic Supply.[41] These stocks include sugar, rice, edible oils, coffee, cereals, and animal feed, designed to last 2–4 months (although may increase to 12months).[42] Although owned by private businesses, these reserves are closely coordinated with the government. FONES can also implement import rule adjustments, rationing, and direct food production management during crises. Moreover, citizens are encouraged to maintain a 1-week emergency food and water supply, supported by government brochures and educational campaigns.[43]

- *Canada*
  Canada's approach to food resilience is evolving from a focus on agricultural emergencies to a broader, system-wide strategy. The *2016 Emergency Management Framework for Agriculture*, developed with provincial and territorial governments, originally concentrated on plant and animal disease.[44] However, disruptions caused by COVID-19, civil unrest, and climate events revealed major gaps, prompting a national review to expand planning across the entire food system—from production to retail—aimed at ensuring Canadians' food security. In parallel, the National Strategy for Critical Infrastructure recognises food as a vital sector.[45] Led by Public Safety Canada, it promotes cross-government and industry collaboration but largely excludes civil society. Agriculture and Agri-Food Canada (AAFC) leads on food sector resilience and is working to improve coordination across supply chains, including farm inputs, processors, and retailers. Canada's Food Policy takes a comprehensive approach by emphasising access to safe, nutritious, and culturally appropriate food, alongside sustainability and economic resilience.[46] Its four pillars—food access, sustainable systems, waste reduction, and economic strength—reflect a balanced commitment to social equity and environmental stewardship. The policy's focus on community-based programmes and infrastructure investments empowers local initiatives, supporting diverse populations and reinforcing the food system's adaptability.

- *France*
  France integrates both emergency recovery efforts and long-term structural reforms. In response to COVID-19, the French government launched the France Relance programme, a broad post-pandemic recovery plan that included support for agri-food.[47] However, France's deeper commitment to food resilience stems from the landmark 2014 Law on the Future of Agriculture, Food and Forestry (Law No. 2014-1170).[48] This legislation laid the foundation for hundreds of local and regional initiatives aimed at strengthening civil food resilience. It empowers communities to re-localise food systems, support short supply chains, and build stronger urban–rural links through Territorial Food Projects (PATs), with legal backing under Article 39 of the law.[49] National coordination bodies such as Terres en Villes and initiatives like Les Greniers d'Abondance have flourished, helping communities assess vulnerabilities and build food security. These

networks proved effective during COVID-19, when areas with local food systems fared better. As a result, France Relance increased support for local food projects, growing the number of PATs from 197 to 400.[50] In 2025, France adopted a new National Strategy for Food, Nutrition and Climate (SNANC), which integrates food sovereignty, biodiversity, and climate goals. France's legal approach has helped normalise food resilience as a democratic priority—offering a model the UK could learn from.[51]

- *Germany*
  Germany takes a comprehensive, state-led approach to food resilience, integrating both emergency preparedness and long-term planning. Two ministries play central roles: the Federal Ministry of Food and Agriculture (BMEL) and the Federal Office of Civil Protection and Disaster Assistance (BBK). Citizens are encouraged to stockpile food and water for 10 days, supported by official guides and online tools—such as a detailed calculator for household needs and a downloadable vegetarian option.[52,53] This scientific and nutritional specificity fosters public trust. A law enables BMEL to manage food supply during crises, supported by strategic reserves of grains, pulses, and condensed milk, primarily for urban centres.[54] These reserves ensure basic daily meals and can be activated upon request by federal states. Germany's food system is also framed within broader critical infrastructure protection. The 2024 "*Good Food for Germany*" strategy represents ambitious, forward-thinking efforts to create sustainable, resilient, and equitable food systems tailored to their national contexts.[55,56] Germany's strategy focuses on strengthening regional supply chains, promoting environmentally friendly consumption, and reducing food waste, all aimed at fostering a more sustainable and locally connected food system. By encouraging support for local producers and shortening supply chains, the plan enhances both environmental sustainability and economic vitality in regional communities. This holistic approach addresses pressing challenges such as climate change and nutrition while laying the groundwork for a future-proof food system.
- *Netherlands*
  The Netherlands lacks a dedicated national food resilience strategy but has long integrated food, health, and environmental goals into policy. With a strong agricultural base and a history of food insecurity, the Dutch government has periodically assessed national food supply risks, notably in a 2013 study modelling crisis scenarios.[57] Though food resilience wasn't central, it recommended stockpiling and a shift to self-sufficiency. Recent developments—including climate threats, the Ukraine war, and geopolitical tensions—have revived food security concerns. Public campaigns like *Denk Vooruit* now advise citizens to prepare for 48 hours post-disaster, yet uptake is low.[58] Food provision is covered under national emergency frameworks, notably through 25 "Security Regions".[59] A new focus on "resilience and resistance" emerged in 2024 amid terrorism warnings and

public broadcast simulations.[60] While food is being reconsidered as critical infrastructure, much of the momentum remains top-down, with limited local strategies or strong public engagement to date.

- *USA*
  The US approach to food preparedness centres on the Department of Homeland Security (DHS) and the Department of Agriculture (USDA), with a primary focus on securing supply chains against threats such as terrorism, natural disasters, and system disruptions. Homeland Security Presidential Directive 9, issued in 2004, and its 2022 update, National Security Memorandum-16, emphasised safeguarding agricultural infrastructure and food systems but placed limited attention on civil food resilience.[61,62] The USDA has identified key vulnerabilities—such as concentrated production, labour shortages, climate change impacts, and transport bottlenecks.[63] Federal Emergency Management Agency (FEMA), incorporated into DHS after 2001, supports states during emergencies but maintains only limited regional food stockpiles. The Strategic National Stockpile holds medical supplies, not food. While DHS recommends households store food and water for several days, public engagement remains uneven, although FEMA promotes preparedness through outreach campaigns like National Preparedness Month.[64]

Other countries have been motivated to create food resilience programmes by a specific and very real threat. Russia's full-scale invasion of Ukraine in 2022 catalysed Latvia's heightened focus on food security, accelerating efforts that had begun after the 2014 annexation of Crimea. Drawing on its geopolitical reality and proximity to Russia, Latvia adopted a whole-of-society approach, formalised in its *2019 National Defence Strategy*.[65] Led by the Ministry of Defence, food is treated as a critical service alongside energy and banking, with ministries assigned distinct roles: agriculture oversees production, and economics manages retail. Key suppliers, known as "K-companies", are bound to support crisis response. Regular crisis simulations and public education—ranging from school card games to multilingual crisis guides—support both institutional coordination and citizen self-reliance. Lithuania has also embedded food resilience within its national security framework, shaped by post-Soviet reforms and EU policy. A legally mandated State Reserve of Agricultural and Food Products is managed by the Ministry of Agriculture and coordinated by the National Crisis Management Center. Like Latvia, Lithuania urges citizens to prepare for 72 hours of self-sufficiency, though public uptake remains low, with only 18% maintaining emergency supplies.[66] While Lithuania has modernised its agricultural sector, food resilience remains largely market-driven, and officials warn that future disruptions may challenge current urban–rural dynamics. Furthermore, Sweden, prompted by Russia's 2014 annexation of Crimea and heightened by the 2022 invasion of Ukraine, focused significantly on addressing the threat from Russia. The 2024 report *Food Preparedness for a New Era* mandated legal obligations for all government levels to ensure food access in crises, with funding and local empowerment to support this.[67] Key agencies—including the Swedish Civil Contingencies Agency and Food Agency—coordinate preparedness through public guidance, contingency planning, and sustainability measures. Citizens are advised to store food for at least 1 week, with information provided via brochures like *If Crisis or War Comes*.[68] Municipalities and businesses also share responsibility for securing supply chains. A 2024 civil defence booklet emphasises psychological readiness and food self-sufficiency, reinforcing food's place in Sweden's national security framework.[69]

A further set of countries, including Italy, placed the responsibility of food resilience on the city. In these instances, the national government does not currently offer public guidance or campaigns on civil food resilience—momentum largely comes from local efforts and engaged communities. Notably, Milan spearheaded the Milan Urban Food Policy Pact (MUFPP) in 2015, a major global initiative involving over 200 cities committed to urban food system action.[70] Although symbolising leadership, this initiative also highlights the national gap: there is no legal requirement for regional or city-level civil food resilience strategies. Across the rest of Italy, Rome and Turin are exploring food rights in local constitutions, with Rome establishing a Food Council.[71] Both cities are actively engaging stakeholders from civil society, businesses, and government to shape inclusive food policies. Turin has launched pilot projects to improve urban agriculture and increase access to healthy, affordable food in underserved neighbourhoods.[72] Meanwhile, Rome's Food Council serves as a platform for coordinating food security initiatives, promoting sustainability, and ensuring citizen participation in decision-making. Outside of Italy, Germany's Munich and Freiburg have developed local strategies to strengthen food system resilience against shocks like climate change and supply disruptions.[73,74] Munich's approach integrates food security into broader climate security discussions, as seen in 2025 at the globally renowned Munich Security Conference, while Freiburg emphasises agroecological methods, with local farmers diversifying production and building social networks to adapt to crises. Both cities highlight the necessity of multi-level coordination—linking local actions with regional and national frameworks—to ensure sustainable, adaptive food systems. These efforts reflect growing recognition that resilient food systems are vital to long-term urban stability and environmental sustainability.

One step up from the city-level, regional governments are considering their food resilience. Swiss cantons, such as Zurich and Berne, actively provide citizens with guidance on emergency stockpiling, including food storage and cooking without electricity.[75] This guidance is part of Switzerland's long-standing civil protection culture. Cantonal and federal authorities regularly update and disseminate practical materials—such as checklists, online calculators, and printed brochures—to ensure households can remain self-sufficient for at least one week during crises.[76] In the US, the focus is state-wide. For example, Maryland and New York have developed their own food resilience strategies.[77] Maryland's strategy includes a formal Food System Resiliency Council, created by legislation in 2021, which addresses supply chain disruptions, emergency food access, and long-term food equity. The state has also integrated food security into broader climate and disaster planning. New York, meanwhile, has established the New York State Emergency Food Plan, which outlines protocols for food storage, distribution, and coordination during crises.[78] It includes partnerships with food banks, emergency feeding programmes, and local governments. Both states emphasise cross-sector collaboration, with agencies coordinating closely with community organisations and private suppliers. These proactive policies aim to build more robust and responsive food systems, especially in the face of climate risks, economic shocks, or large-scale emergencies.

Together, these strategies showcase inspiring models for transforming food systems into sustainable, equitable, and resilient networks that serve both people and the planet. In designing a new food system, there are valuable lessons to learn from much closer to home—from within the UK itself. Scotland has taken bold, forward-thinking legislative steps that offer practical models for embedding food resilience and sustainability into governance. Scotland's Good Food Nation Act adopts a joined-up, rights-based approach to food policy.[79] It requires public bodies to produce food plans and ensures that food is considered across sectors—health, education, environment, and the economy. This coherence strengthens the ability of the food system to respond to shocks and align with public interest goals, including nutrition, climate, and social justice. This legislative framework demonstrates how law can be used not only to manage current pressures but also to future-proof systems. They show that fairness, equity, and resilience can be actively built into governance—providing clear and actionable inspiration for reform.

Notes

1 McKevit, Fraser, Record grocery sales in UK as shoppers prepare for lockdown, Kantar, 2025, https://www.kantar.com/inspiration/fmcg/record-grocery-sales-in-uk-as-shoppers-prepare-for-lockdown/

2 Banton, Caroline, just in time (JIT): Definition, example, pros and cons, Investopedia, 2025, https://www.investopedia.com/terms/j/jit.asp

3 Benton, Tim, & Froggat, Anthony, Food politics and polices in post Brexit Britain, Chatham House, 2019, https://www.chathamhouse.org/2019/01/food-politics-and-policiespost-brexit-britain

4 Wentworth, Jonathan, Rapid response: Effects of COVID-19 on the food supply system, UK Parliament, 2019, https://post.parliament.uk/effects-of-covid-19-on-the-food-supply-system/

5 Wentwoth, Jonathan, Rapid response: Effects of COVID-19 on the food supply system, UK Parliament, 2019, https://post.parliament.uk/effects-of-covid-19-on-the-food-supply-system/

6 Mahanta, Siddhartha, New York's looming food disaster, *Bloomberg*, 2013, https://www.bloomberg.com/news/articles/2013-10-21/new-york-s-looming-food-disaster
7 Patridge, Joanna, & Butler, Sarah, Why are UK supermarkets facing fresh food shortages, *The Guardian*, 2023, Why are UK supermarkets facing fresh food shortages? | Food & drink industry | *The Guardian*.
8 Hansard, UK Food Shortages vol278, UK Parliament, 2023, https://hansard.parliament.uk/Commons/2023-02-23/debates/0814D2CD-836D-4906-BC69-F0F28C047C50/UKFoodShortages
9 Clarke, Jennifer, What are tariffs, how do they work and why is Trump using them, *BBC News*, 2025, https://www.bbc.co.uk/news/articles/cn93e12rypgo
10 Ukraine farm sector indirect losses may reach $83 bin due to Russia invasion, analysts say, *Reuters*, 2024, https://www.reuters.com/markets/commodities/ukraine-farm-sector-indirect-losses-may-reach-83-bln-due-russian-invasion-2024-10-03/
11 Welsh, Caitlyn, & Dodd, Emma, Ukraine decreased grain production creates opportunities for Russia, CSIS, 2025, https://www.csis.org/analysis/ukraines-decreasedproduction-creates-opportunities-Russia
12 Welsh,C.&Glauber,J.,Foodasthe"silentweapon":Russia'sgainsandUkraine'slosses. Center for Strategic and International Studies, Washington DC, USA, 2024, https://www.csis.org/analysis/food-silent-weapon-russias-gains-and-ukraines-losses
13 Ukraine's frontline farms battle Russians and weather, *The Economist*, 2025, https://www.economist.com/europe/2025/07/13/ukraines-front-line-farms-battle-russians-and-weather
14 Cronin, Audrey, Eyai, Hukata, & Shasshank, Joshi, Analysing Israel's strategy in Gaza, *Bookings*, 2025, https://www.brookings.edu/events/analyzing-israels-strategy-in-gaza/
15 UNICEF, Israel must end 'cruel collective punishment' in Gaza, urges UN relief chief, *UN News*, 2025, https://news.un.org/en/story/2025/05/1162806
16 UNICEF, Israel must end 'cruel collective punishment ' in Gaza, urges UN relief chief, *UN News*, 2025, https://news.un.org/en/story/2025/05/1162806
17 Berman, Lazar, IDF says documents show Hamas has been confiscating aid as a matter of policy, *The Times Isreal*, 2025, https://www.timesofisrael.com/idf-saysdocuments-show-hamas-has-been-confiscating-aid-as-a-matter-of-policy/
18 People in Gaza are starving, sick and dying as aid blockade continues, WHO, 2025, https://www.who.int/news/item/12–05–2025-people-in-gaza-starving—s+ick-and-dying-as-aid-blockade-continues
19 Israel's siege now blocks 83% of food aid reaching Gaza, new data reveals, Norwegian Refugee Council, 2024, https://www.nrc.no/news/2024/september/israels-siege-now-blocks-83-of-food-aid-reaching-gaza-new-data-reveals/
20 UN, Global Forum on Food Security and Nutrition (online consultation), FAO, 2016, https://www.fao.org/fsnforum/read_more_about_resilience
21 UN, Adopting an analytical framework on risk and resilience: A proposal for more proactive, coordinated and effective United Nations action, World Food Programme, 2022, https://pmc.ncbi.nlm.nih.gov/articles/PMC9257471/#:~:text=These%20foods%20include%20frozen%20and,-income%20populations%20%2821%29
22 Elaydi, Heather, Food system resilience and land restoration, UN, 2022, https://www.unccd.int/sites/default/files/2022–03/UNCCD%20GLO%20WP%20food.Pdf
23 Nestle reports full year results for 2023, Nestle, 2024, https://www.nestle.com/media/pressreleases/allpressreleases/full-year-results-2023
24 Nestles commitment to sustainability, Nestle, 2023, https://www.nestleusa.com/sustainability
25 BSI Organizational Resilience Report Index 2021, BSI, 2021, https://www.bsigroup.com/LocalFiles/en-US/Whitepapers/Organizational-Resilience/BSI-whitepaper-harnessing-experience-embracing-opportunity.pdf

26 UN, Resilience: Enhanced resilience of people, communities and ecosystems is key to sustainable food and agricultural systems, FAO, 2018, https://www.fao.org/agroecology/knowledge/10elements/balance/en/?page=5&ipp=5&tx_dynalist_pi1%5Bpar%5D=YToxOntzOjE6IkwiO3M6MToiNSI7fQ%3D%3D
27 UN, Resilience: Enhanced resilience of people, communities and ecosystems is key to sustainable food and agricultural systems, FAO, 2018, https://www.fao.org/agroecology/knowledge/10elements/balance/en/?page=5&ipp=5&tx_dynalist_pi1%5Bpar%5D=YToxOntzOjE6IkwiO3M6MToiNSI7fQ%3D%3D
28 Peck, Helen, Resilience in food chain: The study of business continuity management in the food and drink industry, Science Gines, 2006.
29 GOV.UK, The UK government resilience framework, Cabinet Office, 2023, The UK Government Resilience Framework (HTML) – GOV.UK.
30 GOV.UK, National Risk Register 2023, HM Government, 2023, https://assets.publishing.service.gov.uk/media/64ca1dfe19f5622669f3c1b1/2023_NATIONAL_RISK_REGISTER_NRR.pdf
31 GOV.UK, A UK government food strategy for England, considering the wider UK food system, DEFRA, 2025, https://www.gov.uk/government/publications/a-uk-government-food-strategy-for-england/a-uk-government-food-strategy-for-england-considering-the-wider-uk-food-system#ministerial-foreword
32 GOV.UK, A UK government food strategy for England, considering the wider UK food system, DEFRA, 2025, https://www.gov.uk/government/publications/a-uk-government-food-strategy-for-england/a-uk-government-food-strategy-for-england-considering-the-wider-uk-food-system#ministerial-foreword
GOV.UK, A UK government food strategy for England, considering the wider UK food system, DEFRA, 2025, https://www.gov.uk/government/publications/a-uk-government-food-strategy-for-england/a-uk-government-food-strategy-for-england-considering-the-wider-uk-food-system#ministerial-foreword
33 GOV.UK, A UK government food strategy for England, considering the wider UK food system, DEFRA, 2025, https://www.gov.uk/government/publications/a-uk-government-food-strategy-for-england/a-uk-government-food-strategy-for-england-considering-the-wider-uk-food-system#ministerial-foreword
34 GOV.UK, A UK government food strategy for England, considering the wider UK food system, DEFRA, 2025, https://www.gov.uk/government/publications/a-uk-government-food-strategy-for-england/a-uk-government-food-strategy-for-england-considering-the-wider-uk-food-system#ministerial-foreword
35 GOV.UK, A UK government food strategy for England, considering the wider UK food system, DEFRA, 2025, https://www.gov.uk/government/publications/a-uk-government-food-strategy-for-england/a-uk-government-food-strategy-for-england-considering-the-wider-uk-food-system#ministerial-foreword
36 Lang, Tim, & Neumann, Natalie, & So, Anthony, Just in case: 7 Steps to narrow the UK civil food resilience gap, National Preparedness Commission, 2025, https://nationalpreparednesscommission.uk/wp-content/uploads/2025/02/NPC-Just-in-Case-Executive-Summary_PDF-Download.pdf
37 Lang, Tim, & Neumann, Natalie, So, Anthony, Just in case: 7 Steps to narrow the UK civil food resilience gap, National Preparedness Commission, 2025, https://nationalpreparednesscommission.uk/wp-content/uploads/2025/02/NPC-Just-in-Case-Executive-Summary_PDF-Download.pdf
38 Lang, Tim, Neumann, Natalie, & So, Anthony, Just in case: 7 Steps to narrow the UK civil food resilience gap, National Preparedness Commission, 2025, https://nationalpreparednesscommission.uk/wp-content/uploads/2025/02/NPC-Just-in-Case-Executive-Summary_PDF-Download.pdf

39 Lang, Tim, Neumann, Natalie, & So, Anthony, Just in case: 7 Steps to narrow the UK civil food resilience gap, National Preparedness Commission, 2025, https://nationalpreparednesscommission.uk/wp-content/uploads/2025/02/NPC-Just-in-Case-Executive-Summary_PDF-Download.pdf
40 The Federal Assembly of the Swiss Confederation, Federal act on national economic supply, 2025, https://www.fedlex.admin.ch/eli/cc/2017/308/en
41 Lang, Tim, Neumann, Natalie, & So, Anthony, Just in case: 7 Steps to narrow the UK civil food resilience gap, National Preparedness Commission, 2025, https://nationalpreparednesscommission.uk/wp-content/uploads/2025/02/NPC-Just-in-Case-Executive-Summary_PDF-Download.pdf
42 Federal Office for National Economic Supply (FONES), We advise emergency supplies, Swiss Federal Authorities, 2025, https://www.bwl.admin.ch/en/we-advise-emergency-supplies
43 Canada.ca, Emergency Management Framework for Agriculture in Canada, Agriculture Canada, 2016, https://agriculture.canada.ca/sites/default/files/legacy/resources/prod/doc/pdf/emf-gu_2016-eng.pdf
44 Canada.ca, National strategy for critical infrastructure, Public Safety Canada, 2022, https://www.publicsafety.gc.ca/cnt/rsrcs/pblctns/srtg-crtcl-nfrstrctr/index-en.aspx
45 Canada.ca, The food policy for Canada, Government of Canada, 2019, https://agriculture.canada.ca/en/department/initiatives/food-policy-canada
46 GOUV.FR, France relance recovery plan: Building the France of 2030, France Diplomacy, 2020, https://www.diplomatie.gouv.fr/en/french-foreign-policy/economic-diplomacy-foreign-trade/promoting-france-s-attractiveness/france-relance-recovery-plan-building-the-france-of-2030/
47 GOUV.FR, Law No. 2014-1170 of 13 October 2014 on the future of agriculture, food and forestry, Government of the French Republic, 2014, https://www.legifrance.gouv.fr/loda/id/JORFTEXT000029573022
48 GOUV.FR, Evaluation Committee of the National Recovery Plan “France Relance” – Second Report, France Stratégie & Inspection Générale Des Finances, 2022, https://www.strategieplan.gouv.fr/files/files/Publications/English%20Articles/France%20Relance%20-%20Deuxième%20rapport/ang_synthese_france_relance_decembre_2022.pdf
49 Lamine, Claire, Territorial food projects in France, FAO, 2023, https://www.fao.org/family-farming/detail/en/c/1679099
50 Acteurs de la société civile, Stratégie nationale pour l’alimentation, la nutrition et le climat (SNANC), La Confédération Syndicale des Familles (La CSF), 2025, https://www.la-csf.org/wp-content/uploads/2025/04/SNANC-decryptage-VF.pdf
51 The Federal Office of Civil Protection and Disaster Assistance (BBK), Stockpiling, 2022, https://www.bbk.bund.de/EN/Prepare-for-disasters/Personal-Preparedness/Stockpili ng/stockpiling_node.html
52 Bundesministerium für Landwirtschaft, Ernährung und Heimat, Food preparedness, 2024, https://www.ernaehrungsvorsorge.de/english
53 Bundesministerium für Landwirtschaft, Ernährung und Heimat, Food preparedness, 2024, https://www.ernaehrungsvorsorge.de/english
54 Federal Ministry of Agriculture, Food and regional identity, good food for Germany – The Federal Government’s food and nutrition strategy, BMLEH, 2024, https://www.bmleh.de/SharedDocs/Downloads/EN/_Food-and-Nutrition/Food-and-Nutrition-Strategy.pdf
55 Canada.ca, The food policy for Canada, Government of Canada, 2019, https://agriculture.canada.ca/en/department/initiatives/food-policy-canada
56 Terluinm, Ida, et al., Food supply in the Netherlands under exceptional crisis conditions, Wageningen, 2013, https://www.wur.nl/upload_mm/d/d/2/2aee638f-

ea64-446d-9931-a729594c1686_2013-012%20Kamphuis-Terluin_UK%20 SUMMARY.pdf
57 Denk vooruit, Home page, 2025, https://english.denkvooruit.nl
58 Government of the Netherlands, Act of 11 February 2010, containing provisions for the fire services, disaster management, crisis management and medical assistance (Security Regions Act), 2010, https://disasterlaw.ifrc.org/sites/default/files/media/disaster_law/2020-08/dutch-security-regions-act-official-consolidated-version%20 2010.pdf
59 Government.nl, Letter on LNV's international commitment to transboundary food system, Government of the Netherlands, 2024, https://www.government.nl/documents/parliamentary-documents/2024/03/15/letter-on-lnvs-international-commitment-to-resilient-transboundary-food-systems
60 US.GOV, Homeland security presidential directive/HSPD 9 – Defence of United States Agriculture and Food, GPO, 2004, https://www.govinfo.gov/content/pkg/PPP-2004book1/pdf/PPP-2004-book1-doc-pg173.pdf
61 WH.GOV, National security memorandum on strengthening the security and resilience of United States Food and Agriculture, The White House, 2022, https://irp.fas.org/offdocs/nsm/nsm-16.pdf
62 US.GOV, Household food security in the United States in 2023, USDA, 2024, https://www.ers.usda.gov/publications/pub-details?pubid=109895
63 Ready.gov, Suggested emergency food supplies. In: US Department of Homeland Security website. 2022, https://www.ready.gov/food
64 Bērziņš, J., Latvia's comprehensive defence approach. *Bezpieczeństwo. Teoria I Praktyka, XLVIII* (3), 223–235, 2022, DOI:10.48269/2451-0718-btip-2022-3-016
65 Lithuanian launches major campaign to boost emergency preparedness, *Euro News*, 2023, https://www.euronews.com/2023/01/11/lithuania-launches-major-campaign-to-boost-emergency-preparedness
66 Prime Minister hosted a meeting of the Nordic Prime Ministers and heads of Government, Government Offices of Sweden, 2025, https://www.government.se/press-releases/2025/10/prime-minister-to-host-meeting-of-nordic-prime-ministers-and-heads-of-government/
67 Swedish Civil Contingencies Agency, If crisis or war comes, 2018, https://www.jonkoping.se/download/18.6a087b4c169dcff8c903056e/1556627549015/Engelska.pdf
68 Swedish Civil Contingencies Agency, In case of crisis or war, 2024, https://rib.msb.se/filer/pdf/30874.pdf
69 Milan Urban Food Policy Pact, 2015, https://www.milanurbanfoodpolicypact.org
70 Food Policy Councils, Home page, 2025, https://www.fpc2025.com
71 European projects for the city of Torino, 2024, https://www.torinoeuprojects.it/en/cofarm4cities/
72 Food future for Freiburg, 2025, https://ernaehrungsrat-freiburg.de/f4/
73 Schneider, Hendrik, Munich could achieve regional food sufficiency, Research in Germany, 2023, https://www.research-in-germany.org/idw-news/en_US/2023/8/2023–08-08_Munich_could_achieve_regional_food_self-sufficiency.html
74 Niner, Huber, What to keep in your emergency store, Migros, 2024, https://corporate.migros.ch/en/story/emergency-supplies
75 Swiss Federal Authorities, We advise emergency supplies, 2025, https://www.bwl.admin.ch/en/we-advise-emergency-supplies
76 Maryland.gov, Resilient food systems infrastructure program, Maryland Government, 2024, https://mda.maryland.gov/maryland_products/Pages/Resilient-Food-Systems-Infrastructure-Program.aspx
77 Ny.gov, The emergency food assistance program, Office of the General Services, 2023, https://ogs.ny.gov/usda-food-distribution/emergency-food-assistance-program-tefap
78 GOV.UK, Good Food Nation Scotland Act 2022, legislation.gov.uk, 2022, https://www.legislation.gov.uk/asp/2022/5/contents

3 Multi-faceted food resilience

The Government has taken an important and commendable first step in recognising the urgent need for supply chain resilience by making it a core objective of the National Food Strategy 2025 and in the National Preparedness Commission commissioning Professor Lang to look at critical food infrastructure. This is a critical acknowledgement of just how exposed the UK's food system has become to external shocks—from climate change and global conflicts to economic volatility and market concentration. However, while this recognition marks real progress, supply chain resilience represents only one strand of a much larger web of food system resilience. If the Government focuses too narrowly on logistics and trade, it risks addressing the symptoms of instability without tackling its deeper structural causes. To deliver genuine resilience, food policy must consider the interconnections between production, supply, access, and nutrition. The resilience of supply chains depends fundamentally on the resilience of the farmers who produce our food, the infrastructure that supports them, and the consumers who rely on affordable, healthy options. Climate-resilient farming, fair market access, and strong local food networks all underpin the very stability the Government seeks to achieve. When one link in this chain fails—be it through extreme weather damaging harvests, price pressures driving farmers out of business, or households unable to afford nutritious food—the entire system becomes more fragile. Building resilience into the food system must therefore be intentional, proactive, and systemic. It means investing in long-term food security through diversified supply chains, sustainable domestic production, and robust local food networks that can respond quickly to disruption. It means supporting fairer markets that reward environmental stewardship and ensure farmers receive prices that reflect the true cost of sustainable production. And it means ensuring that every citizen has access to affordable, nutritious food, even in times of crisis. By recognising that supply chain resilience alone cannot exist without environmental, social, and economic resilience, the UK can move from a reactive model—one that merely copes with shocks—to a proactive one that anticipates, adapts, and thrives.

DOI: 10.4324/9781003581871-4

A more holistic approach to resilience should therefore be adopted by the Government and integrated through all its food and farming policy, starting with the National Food Strategy Action Plan that will lay out a roadmap to deliver the pledges within the National Food Strategy. The Action Plan is due in 2026. By extending the understanding of food resilience across six key dimensions (environmental, economic, social, supply chain, health, and geopolitical), true food resilience can begin to be realised. The 2025 National Food Strategy Action Plan presents a real opportunity to embed these interlinked dimensions of resilience into UK food policy. This will require cross-departmental coordination between the Department for Environment, Food, and Rural Affairs, Department of Health and Social Care (DHSC), the Ministry of Housin, Communites and Local Governance (MHCLG), and the Department for Business, Energy, and Industrial Strategy, breaking down long-standing silos between agricultural, health, and economic policy. The Government should ensure that food and farming are treated as an integrated system, rather than separate policy areas. Farming determines not only what food is produced but also how healthy, affordable, and sustainable that food is. Food policy that ignores how food is grown risks perpetuating diets high in calories but low in nutrition; agricultural policy that ignores public health can lead to overproduction, waste, and environmental harm.

Let's consider the six sub-categories that underpin food resilience in greater depth. They are:

Climate resilience

As we've discussed, the key to the future of food is tackling the climate crisis head-on. Food resilience rests on climate resilience. Climate change is no longer a distant threat; it is reshaping our environment, economy, and food system today. Human activities—burning fossil fuels, deforestation, and intensive farming—have driven global warming, resulting in more frequent and severe floods, droughts, and heatwaves. Over the last 50 years, weather-related disasters have increased fivefold.[1] By 2030, global temperatures are projected to rise to almost 2°C above pre-industrial levels, a tipping point with grave implications.[2] For the UK, resilience in food and farming policy is not optional—it is essential. The UK is already feeling the strain. Since 1900, sea levels have risen nearly 20 cm, with most of that in the past 30 years.[3] Heatwaves, once rare, now claim thousands of lives, while flooding threatens millions of homes. In 2022, record heat caused severe drought, reducing poultry yields by 9% and leaving livestock farmers short of forage.[4] Climate shocks like these endanger food security, expose weaknesses in supply chains, and push up prices. Building resilience into farming policy is therefore critical to protect both farmers and consumers.

At the same time, the UK faces a biodiversity crisis that compounds climate risks. According to the *2023 State of Nature* report, nearly one in six species is at risk of extinction, and monitored populations have declined by 19% since 1970.[5] Agriculture is central to this story. The conversion of wildflower meadows, woodlands, and wetlands has erased habitats, while pesticides and fertilisers have damaged pollinator populations and soil biodiversity. Because three-quarters of food crops depend on pollination, these losses weaken farming's resilience just as climate pressures intensify.[6] Supporting nature-friendly farming is therefore a resilience strategy, not a luxury. Globally, the picture is similar. The WWF's *Living Planet Report 2024* shows a 73% decline in wildlife populations over the last half-century, and the UN estimates 1 million plant and animal species are at risk.[7,8] The decline of pollinators alone could lead to over 500,000 premature deaths annually due to reduced access to nutritious food.[9] These trends reveal how climate and ecological breakdowns undermine the very foundations of farming. A resilient food system cannot be built without restoring the ecosystems it depends on.

Yet climate and biodiversity are not the only environmental concerns shaping the future of food. The Planetary Boundaries framework, developed by the Stockholm Resilience Centre, underscores the scale and interconnectedness of the environmental challenges we face.[10] It identifies nine biophysical limits—including climate change, land-use conversion, freshwater use, nitrogen and phosphorus pollution, and loss of biosphere integrity—that together maintain the stability of the Earth system on which all human activity depends. Alarming evidence shows that we are already exceeding several of these boundaries, placing global food security and livelihoods at growing risk. In particular, the overuse of synthetic nitrogen driven by the Green Revolution, fertilisers has severely disrupted the nitrogen cycle, driving soil degradation, contaminating waterways, and fuelling oceanic dead zones. It also releases nitrous oxide, a greenhouse gas roughly 270 times more potent than carbon dioxide, intensifying climate instability.[11] These impacts ripple through ecosystems and economies alike, undermining the very foundation of sustainable food production. If we are to build a climate-resilient and nature-positive food system, we must look beyond carbon metrics and actively consider how each of the Planetary Boundaries interacts with agriculture. Only by operating within these ecological limits can farming truly sustain both people and the planet for generations to come.

Yet, we need to think of the climate impact outside our shores. The UK imports nearly half of its food, including most fruit and vegetables.[12] Firstly, this exposes our food security to wider climate impacts. This dependence leaves us exposed to climate shocks overseas. Southern European droughts reduce vegetable exports; Asian floods disrupt rice and spices; and in 2021, Spain's Storm Filomena left UK supermarkets rationing salad vegetables.[13,14,15] These global extreme weather disruptions are set to become more common. In the last 50 years in Europe, a forest area more than twice the size of Portugal has been destroyed by fires, and more than 246,000 people have lost their lives due to extreme climate-related weather events.[16,17] Beyond this vulnerability, the dependence on imports risks offshoring the environmental damage of our food system. We might strengthen our environmental regulations here, but if we do little to set a comparable bar on the environmental impact of the food we import, we are just as culpable. Our global climate is just as connected as our global food system is. We have a responsibility as a Global North nation that is buying food from every corner of the globe to make sure that our demand is not driving environmental degradation in other communities. The most obvious examples of this are the use of unsustainable palm oil extracted from virgin rainforests and the import of soy for animal feed that drives Amazonian deforestation. Palm oil and soy, which are in the category of 'oilseeds' together with smaller commodities such as sunflower, rapeseed, and sesame, drive 18% of deforestation in the Amazon.[18] These are mostly consumed by the Global North. For instance, the EU is the second largest importer of soy after China, severely outsourcing the environmental impact of its food consumption. Tropical deforestation is, in fact, estimated to account for 1/6 of the carbon footprint of a European citizen's diet.[19] There are some signs that progress is possible. In 2023, the EU adopted a new regulation on "deforestation-free products" covering seven commodities (soy, cattle, palm oil, cocoa, coffee, rubber and wood) and their derived products.[20] Under the regulation, the actors in the supply chain (e.g. operators or traders) must demonstrate that the products placed on the European market do not originate from recently deforested land. This is an important step in the right direction, but it is still crucial to always consider that climate resilience in global food production must be about fairness, since climate change falls the hardest on vulnerable Global South producers, whether through flooding, drought, or desertification.

Genetic diversity

The global food system rests on a surprisingly narrow base of crop diversity, a dynamic that creates vulnerability. Out of roughly 250,000 plant varieties available for agriculture, fewer than 3% are in use. In practice, the world relies on about 150 species, with just 12 crops accounting for three-quarters of global food consumption.[21] Rice, wheat, and maize alone provide more than half of global calories, joined by soy, potatoes, and sugar to dominate food energy.[22] This concentration may appear efficient, but it poses serious risks. Much of this production takes place in monocultures, where a single crop variety is cultivated across vast areas. While monocultures simplify mechanisation and trade, their genetic uniformity makes them inherently fragile. A pest or pathogen that overcomes one plant's defences can spread unchecked, wiping out entire harvests. The consequences are clear: reliance on a small number of crops and varieties concentrates risk. A single blight could wipe out an entire crop and severely limit supply. The Irish Potato Famine in the 1840s, when reliance on a single variety (the Lumper) left millions vulnerable to blight, is a stark historical reminder. The banana industry has experienced similar crises, with the once-dominant Gros Michel wiped out by Panama disease and today's Cavendish at risk from a new fungal strain. In natural ecosystems, genetic diversity acts as a buffer: different species and varieties respond differently to threats, limiting the scale of damage. Monocultures, in contrast, offer pests a stable, uniform food source and often depend on chemical inputs to suppress threats. Yet pests can evolve resistance, leading to escalating input use and further ecological degradation. Soils depleted by intensive farming further weaken plants' resilience. This fragility is compounded by climate change, which is enabling pests and diseases to spread into new regions, exposing already vulnerable monocultures to additional shocks. Policy must step up and incentivise diversity through farm support schemes, funding farmers who grow heritage varieties, landraces, and diverse rotations.

The narrowing of crop diversity is partly the outcome of breeding priorities. Throughout the 20th century, plant breeding focused on maximising yield and creating uniform, marketable produce. Uniformity made sense for mechanisation, supermarket aesthetics, and global trade, but it came at a cost. Crops bred for yield often lose resilience traits found in traditional landraces. They also lost nutritional density. A University of Texas study comparing USDA data from 1950 to 1999 found notable declines in protein, calcium, iron, and vitamins across 43 vegetables. Similar nutrient declines have been observed in wheat.[23] These reductions, linked to the narrow parameters for breeding, are compounded by declining yields due to soil depletion and erosion. This has real public health impacts: micronutrient deficiencies contribute to impaired immune function, increasing the risk of infectious diseases, and poor growth and cognitive development in children.[24] Even within major crops, genetic diversity has been limited by breeding programmes. Take avocados: of hundreds of cultivars, over 80% of the global supply comes from Hass, prized for its durability in shipping.[25] Apples show the same trend. In the 19th century, the UK cultivated over 2,000 varieties. Today, supermarket shelves are dominated by four: Royal Gala, Braeburn, Jazz, and Pink Lady.[26] Their shelf life and appearance suit supermarket demands, but their dominance has crowded out heritage varieties adapted to local conditions. Animal agriculture is no different. The global commercial broiler chicken breed, Cobb 500, makes up around half of all farmed chickens.[27] Likewise, the dairy cattle breed Holstein shows stark genetic narrowing: for example, one study found an inbreeding rate of ~0.55% per year for Holstein bulls in France.[28] Such narrowing heightens vulnerability to pests and diseases, and erodes cultural food traditions.

The rise of genetically modified and gene-edited crops has not reversed this trend. In fact, it has reinforced the dominance of a handful of mega-crops, particularly soy and maize. While marketed as 'climate-smart', most new breeding techniques still focus on yield and aim to maximise productivity. Just as Borlaug focused on breeding to improve yields and ensure food security, plant genetics remains a key focus of delivering food security. The UK has taken significant steps to embrace new crop breeding techniques through the enactment of the Genetic Technology (Precision Breeding) Act 2023.[29] This legislation, passed in March 2023, establishes a regulatory framework that distinguishes precision-bred organisms (PBOs) from genetically modified organisms (GMOs), thereby facilitating the development and commercialisation of crops with enhanced traits. Under the new law, crops developed using gene editing technologies—such as CRISPR—are subject to a streamlined regulatory process, provided the genetic changes could have occurred naturally or through traditional breeding methods. Similarly, the UK Government has recognised the potential for cell-cultured products (colloquially known as lab-grown meat) to deliver food security in the future. As a result, the UK's Food Standards Agency (FSA) has initiated a pioneering regulatory "sandbox" programme to facilitate the safe introduction of cell-cultivated products (CCPs) into the market.[30] Launched by the FSA in collaboration with Food Standards Scotland (FSS), this 2-year initiative, supported by a £1.6 million grant from the Department of Science, Innovation and Technology's Engineering Biology Sandbox Fund, aims to ensure consumer safety while fostering innovation in the sector. The programme involves collaboration with academic institutions, industry stakeholders, and trade organisations to gather rigorous scientific evidence on CCPs and their production processes. This data will inform regulatory decisions, streamline application assessments, and provide clearer guidance to businesses. The FSA anticipates completing the safety assessment of at least two CCPs within the programme's duration. The sandbox programme aims to accelerate the approval process for CCPs, positioning the UK as a leader in sustainable food innovation.

The trend towards the new plant breeding techniques reinforces monocultures rather than broadening genetic diversity that would help build resilience. Gene editing may one day deliver resilience traits, but the current system of breeding undermines this future potential. The rise of corporate control in agricultural breeding has concentrated power over seeds and animal genetics into the hands of a few multinational companies, limiting farmer autonomy and biodiversity. Through mergers and acquisitions, firms such as Bayer-Monsanto, Corteva, and Syngenta dominate global seed markets and increasingly control livestock genetics. Intellectual property rights—especially patents and plant variety protections—allow these corporations to restrict the use, saving, or sharing of genetic materials, undermining traditional breeding and public research. This commodification of genetic resources narrows genetic diversity, increases input dependency, and threatens the resilience and sovereignty of food systems worldwide. Without systemic change, this ownership of genetics risks deepening dependence on narrow genetic pools and the corporatisation of our food system. A resilient food system cannot be built on a handful of uniform crops. Diversity is our insurance policy in an uncertain world. Without it, our food system remains dangerously exposed.

Economic viability

It must be economically viable for farmers to produce food. Without the ability to sell their produce at a fair price and make a reliable profit, many will simply leave the industry. This is already happening in the UK, where small- and medium-sized farms are under severe pressure. Farmers face an increasingly hostile commercial environment, with supermarket contracts often altered or dropped at short notice and grocery prices kept artificially low. Farmers and producers have long highlighted the problem, which led to the establishment of the Groceries Code Adjudicator. The Grocery Code Adjudicator in the UK was established to address unfair supermarket practices, but its powers remain limited, and many producers still face late payments, contract uncertainty, and prices below production cost.

These challenges are compounded by sharply rising input costs. Energy, fertiliser, and feed prices have escalated in recent years—exacerbated by the war in Ukraine and conflict in the Middle East region, global market volatility, and the UK's dependence on imported gas.[31] Farmers are often "price takers" in markets dominated by a few multinational suppliers of seed, agrochemicals, and machinery, which leaves them unable to negotiate lower costs. Smaller farms, in particular, struggle to absorb these expenses, and many now operate at a loss. Government statistics show that around 30% of farms in England are loss-making, with a further quarter earning less than £25,000 annually.[32] Such margins leave little room to reinvest in infrastructure, improve efficiency, or adopt sustainable practices.

Policy uncertainty also undermines viability. The phasing out of EU subsidies and the rollout of new domestic support schemes such as the such as the ELM scheme have left many farmers uncertain about future income. The administrative burden of applying for grants or participating in pilots has fallen heavily on smaller farms, which often lack the capacity to navigate complex schemes. In parallel, labour shortages following Brexit have increased wage costs, particularly for horticulture, while insurance premiums and borrowing costs have risen sharply. These pressures create a vicious cycle: with reduced capacity to plan and invest, farmers become more exposed to shocks—from extreme weather to market volatility. Without structural reforms and targeted support, more will exit the sector, eroding rural economies and weakening the UK's long-term food security.

On the other side of the economic viability argument is the fact that food must also be affordable for consumers. Farmers cannot thrive if citizens are priced out of healthy and nutritious food, and where there is always retail pressure for cheaper and cheaper food. Yet in the UK today, millions of families struggle to access the diets that public health guidance recommends.[33] The problem is not simply about individual choices but structural barriers: the price of healthy food has risen more quickly than the price of processed alternatives, while wages and welfare payments have not kept pace with the cost of living. This affordability gap has clear health consequences. As discussed, poor diets are now one of the leading drivers of ill health in the UK, contributing to rising levels of obesity, type 2 diabetes, and cardiovascular disease, resulting in the NHS spending billions each year treating diet-related disease known as NCDs.[34,35] Research shows that children in the most deprived areas are more than twice as likely to be obese as those in the least deprived areas.[36] These inequalities emerge early in life and persist into adulthood, driving the burden of non-communicable disease.

This price dynamic drives demand for cheap, ultra-processed foods that are predominantly produced cheaply (using agrochemicals) and sold by large food conglomerates. These products dominate supermarket shelves because they can be manufactured and distributed at scale, while fresh, seasonal, and sustainably grown produce remains comparatively expensive. Small-scale producers, particularly horticulturalists and mixed farmers, struggle to compete within a marketplace that rewards volume over value. As input costs rise and margins shrink, many are forced out of business or into reliance on low-value wholesale contracts. It is citizens who lose out, doomed to be trapped in poverty by cycles of ill dietary health, pushed towards calorie-dense but nutrient-poor options. The cost-of-living crisis has only magnified these inequalities. Unless food affordability and fairness are placed at the centre of agricultural and economic policy, both producers and consumers will continue to lose out.

Supply chain adaptation

The resilience of the UK food system depends on increasing the diversity of our routes to market. Relying on a single pathway for moving food is inherently fragile because when that route falters, the entire supply chain can break down. This fragility was starkly revealed during the COVID-19 pandemic, when ports and borders closed and supermarket shelves emptied almost overnight. Panic buying exposed how finely balanced and overstretched our just-in-time model had become, with little flexibility to absorb shocks. Farmers supplying hospitality and catering suddenly found themselves without buyers, while retailers struggled to redirect produce to new markets. The collapse of food service supply chains revealed how little redundancy existed within the system—how dependent we had become on centralised distribution networks, long-haul transport, and narrow commercial relationships. At the same time, many small, local producers demonstrated remarkable adaptability, developing new routes to market almost instantly direct delivery schemes, veg boxes, and local collection hubs that kept communities fed when large-scale systems faltered. These examples underscore that diversity, decentralisation, and local infrastructure are vital ingredients of resilience. COVID showed that resilience is not about efficiency at scale, but about flexibility and connection—about ensuring that multiple, overlapping pathways exist between producers and consumers so that, when disruption comes, the system can bend without breaking. Globally, the food system rests on the shipping industry, with containers criss-crossing the oceans in tightly coordinated patterns. On the surface, this appears resilient, but in reality, it is deceptively fragile, since the entire system is reliant on key access points. Strategic locations like the Suez and Panama Canals are choke points that keep global food flowing; when they are blocked, as in 2021 when the Ever Given lodged in the Suez Canal, or the 2026 blockage of the Straits of Hormuz by Iran, cascading shortages ripple across markets. Future risks are even more concerning: rising tensions in the Taiwan Strait or further conflict in the Gulf of Aden could make key sea lanes inaccessible. Climate change compounds these risks, with falling water levels in the Panama Canal and intensifying storms in the South China Sea threatening the routes that keep UK supermarket shelves stocked. With over 80% of global goods moving by sea, any disruption can send shockwaves across supply chains.[37]

Fragility, however, is not limited to the global scale. Domestically, the UK has built a food system that is overly dependent on supermarkets. Over recent decades, smaller farms, wholesalers, and retailers have been progressively sidelined as supermarkets consolidated their dominance. Today, four major retailers—Tesco, Sainsbury's, Asda, and Morrisons—control around 70% of the grocery market.[38] This concentration gives them enormous power to dictate terms to suppliers, often through restrictive contracts, downward price pressure, and limited choice for consumers. Farmers frequently face sudden changes to contracts, late payments, or prices that barely cover production costs. The dominance of supermarkets creates a bottleneck in the food system every bit as restrictive as a shipping canal—producers must pass through them to reach consumers, and if that channel weakens or collapses, the entire system is destabilised. The early months of COVID-19 showed a glimpse of an alternative. With international trade disrupted and supermarket supply chains under strain, small producers developed new routes to market almost overnight: direct-to-consumer deliveries, local box schemes, and community-based food hubs flourished. These innovations demonstrated how quickly the food system can adapt under pressure. Yet, as the pandemic wore on, supermarket supply chains reasserted themselves, and much of that innovation was lost. Some initiatives survived, such as Butter and Crust, but others, like FarmDrop, faltered—showing both the potential and precarity of building alternatives outside the supermarket model.

For UK food supply to be resilient, supply chains must be adaptive and diverse. Dependence on a single dominant route—whether international shipping lanes or supermarket contracts—creates systemic fragility. True resilience requires multiple, complementary channels that allow food to flow from farm to fork through a variety of pathways. It also requires policies that challenge concentration and rebalance power away from a handful of retailers towards a more plural and resilient marketplace. Supply chain resilience, from a food systems perspective, must be defined as the ability to anticipate, prepare for, respond to, and recover from disruptions while maintaining continuous access to safe, nutritious, and affordable food. Without diversifying our routes to market—globally and domestically—the UK will remain dangerously exposed to external shocks and internal bottlenecks. Building resilience means breaking dependence on both fragile shipping choke points and the concentrated dominance of supermarkets, and creating a food system with enough flexibility to withstand the crises that lie ahead.

Market competition

Beyond domestic supermarket concentration, the global consolidation of food and farming companies creates an even greater challenge for food stability. Over recent decades, a series of mega-mergers has concentrated enormous power in the hands of a few multinational corporations. In the farming sector, the mergers of Bayer with Monsanto, Syngenta with ChemChina, and the dominance of companies like Corteva and BASF now mean that just a handful of firms control up to 85% of the global seed market and nearly 80% of the agrochemical market.[39] Fertiliser production is similarly consolidated, with Nutrien, Yara International, CF Industries, and Mosaic together controlling over half the global market.[40] This concentration leaves farmers with little ability to shop around for inputs: prices are effectively dictated by a few dominant suppliers, and spikes in energy or fertiliser costs—as seen following Russia's invasion of Ukraine—can devastate farm budgets. The consolidation trend extends beyond farming inputs to food manufacturing and retail. Global giants like Nestlé, Kraft Heinz, Unilever, and Kellogg's control vast portions of the food market. This double squeeze—high input costs dictated by agribusiness multinationals, and low output prices dictated by retail oligopolies—leaves farmers caught in the middle, often unable to secure a fair return.

Unless governments act to address this imbalance of power, the viability of family and small-scale farms will continue to decline, paving the way for further consolidation by corporate agribusinesses. If unchecked, the future of farming risks becoming a patchwork of corporate-run industrial farms, detached from rural communities and local economies. To avoid this outcome, governments must strengthen competition policy, revisit merger and acquisition rules, and provide targeted support for small and medium-sized food and farming enterprises. Without action, resilience in the UK's food system will remain dangerously compromised.

Geopolitical turmoil

Pen Vogler captured the precariousness of modern food systems in *Stuffed: A Political History of What We Eat and Why It Matters*: "If you want to really scare yourself, think how your shopping basket might look in the midst of a full-scale war, a nuclear attack, cyber-warfare, an energy failure; a hostile nation embedded in our economy that attempts to restrict our access to food and power".[41] This warning is no longer speculative. The world is increasingly unstable, and food systems—because they are globalised and interconnected—are acutely vulnerable. Conflict today is fought not only with weapons but also with trade, tariffs, cyberattacks, and the strategic weaponisation of food. This fragility is intensified by global trade dependency. Around a quarter of the world's food crosses borders.[42] The UK is especially exposed: nearly half of our food is imported, including the majority of fruit and vegetables.[43] Once, Britain could assume privileged access to European surpluses, but as Professor Tim Lang warns, "The UK cannot expect to have first call on regional surpluses … nor can we make glib assumptions that such surpluses will continue to exist or be priority shipping to the UK. Other countries now have deep pockets and more powerful diplomatic and military influence".[44]

The balance of food power is shifting. Nations in the Global South are expanding capacity, modernising production, and becoming both major consumers and exporters. China, in particular, has invested heavily in agriculture and infrastructure across Africa, Asia, and Latin America, securing long-term food access while using its purchasing power to reshape trade flows. The ongoing US–China trade war, China has quickly pivoted soybean imports away from America towards Brazil and domestic production—leaving American farmers devastated but also limiting supply for other nations. This demonstrates how a single political dispute can reorder global markets overnight. The UK, now outside the EU, is negotiating alternative trade agreements in a context of increasing protectionism. Populist leaders worldwide, from the USA to India and Turkey, have embraced tariffs and economic nationalism. These policies inflate food prices, destabilise markets, and reduce the predictability of supply. President Trump's trade war with China in 2025 immediately saw tariffs ripple through global agricultural markets, threatening both affordability and availability. America, under President Trump, is increasingly using tariffs on food and other agricultural products to weaken the economies of other nations, in particular China. Such moves underline that the era of "ever-freer trade" is over.

In earlier sections, we have discussed how the last decade has exposed the vulnerabilities within our current food system and the lack of resilience our supply chains have to global shocks. The COVID-19 pandemic, the war in Ukraine, and the ongoing conflicts in the Middle East all demonstrate the fragility of our supply chains. Ukraine's role as a major exporter of wheat, barley, and sunflower oil meant that the Russian invasion in 2022 sent shockwaves across global markets, driving up prices and limiting supply. Recognising its power, the Russian leadership weaponised food with Russian drones targeting farmland directly.[45,46] Similarly, in Gaza, Israeli forces have curtailed humanitarian supplies and food imports, which they have justified saying that the goal was to cut off the resources of Hamas.[47] These actions show how food access is increasingly being deployed as a tool of war. These are not isolated examples—they signal a new era in which our globalised food system is a lever of geopolitical strategy.[48]

These conflicts and tactics are not new. They have existed through history. However, for much of history, wars and famines were devastating but localised. Today, because our food system is so globalised, conflicts thousands of miles away can directly affect UK consumers. The lesson is stark: food security is no longer simply a question of supply and demand, but one of geopolitics. Conflict, tariffs, and global power shifts can starve supply chains as surely as droughts or pests. The UK cannot afford complacency. By embedding resilience—through diversity, domestic production, and strategic planning—food policy can move from reactive crisis management to proactive security. The unsettling vision of empty shopping baskets that Vogler sketched may seem extreme, but if instability deepens, it is not implausible. The only way to ensure it does not become reality is to plan for resilience now.

Now, let's consider how to deliver against each of these categories in greater detail….

Notes

1 World Meteorological Organization, Weather-related disasters increase over past 50 years, causing more damage but fewer deaths, WMO, 2021, https://wmo.int/media/news/weather-related-disasters-increase-over-past-50-years-causing-more-damage-fewer-deaths
2 World Meteorological Organization, WMO global annual to decadal climate update 2025–2029, WMO, 2025, https://wmo.int/sites/default/files/2025-05/WMO_GADCU_2025–2029_Final.pdf
3 European Environment Agency, Global and European sea level rise, EEA, 2025, https://www.eea.europa.eu/en/analysis/indicators/global-and-european-sea-level-rise
4 Barker, L. J., et al., An appraisal of the severity of the 2022 drought and its impacts. *Weather*, *79* (7), 208–219, 2024, DOI:10.1002/wea.4531
5 State of Nature Partnership, State of Nature Report, 2023, https://stateofnature.org.uk
6 UN, Pollinators vital to food production worldwide, FAO, *News Release*, 2016, https://www.fao.org/newsroom/detail/Pollinators-vital-to-our-food-supply-under-threat/en
7 WWF, Living Planet Report 2024 – A system in Peril, WWF, 2024, https://wwflpr.awsassets.panda.org/downloads/2024-living-planet-report-a-system-in-peril.pdf
8 UN, Nature's dangerous decline 'unprecedented'; species extinction rates 'accelerating', UN, 2024, https://www.un.org/sustainabledevelopment/blog/2019/05/nature-decline-unprecedented-report/
9 Carrington, D., Global pollinator losses causing 500,000 early deaths a year – Study, *The Guardian*, 2023, https://www.theguardian.com/environment/2023/jan/09/global-pollinator-losses-causing-500000-early-deaths-a-year-study

10 Stockholm Resilience Centre, Planetary boundaries, 2025, https://www.stockholmresilience.org/research/planetary-boundaries.html
11 UN, Rise in nitrous oxide emissions endangers pathway to 1.5°C, the ozone layer, and human health, UNEP, 2024, https://www.unep.org/news-and-stories/press-release/rise-nitrous-oxide-emissions-endangers-pathway-15degc-ozone-layer
12 GOV.UK, UK food supply sources. In: *United Kingdom Food Security Report 2021*. DEFRA, 2021, https://www.gov.uk/government/statistics/united-kingdom-food-security-report-2021/united-kingdom-food-security-report-2021-theme-2-uk-food-supply-sources
13 Joint Research Centre, Summer drought reduces yields in south-east Europe, European Commission, 2025, https://joint-research-centre.ec.europa.eu/jrc-news-and-updates/summer-drought-reduces-yields-south-east-europe-2025-09-22_en
14 Venkatappa, M. et al., Impacts of droughts and floods on croplands and crop production in Southeast Asia – An application of Google Earth Engine. *Science of the Total Environment, 795* (170191), 2021, DOI:10.1016/j.scitotenv.2021.148829
15 Wood, Z., Spanish snowstorm leads to UK salad shortage, *The Guardian*, 2021, https://www.theguardian.com/business/2021/jan/15/spanish-snow-storm-uk-salad-shortage-lettuce-tomato
16 Council of the European Union, Climate change costs lives and money, European Union, 2022, https://www.consilium.europa.eu/en/infographics/climate-costs/#:~:text=The%20financial%20losses%20caused%20by%20extreme%20weather%20and,was%20the%20highest%20for%20Germany%2C%20Italy%20and%20France.
17 European Environment Agency, Total fatalities by event type for different country groups, EEA, 2025, https://www.eea.europa.eu/en/analysis/maps-and-charts/total-fatalities-by-event-type-for?activeTab=a7caf3b5-7254-4a24-8919-693d4115158b
18 Ritchie, A., Drivers of deforestation, a world in data, 2021, https://ourworldindata.org/drivers-of-deforestation#article-citation
19 Joint Research Centre, The Amazon region in 2022 and 2023: Deforestation, forest degradation and the risk of growing soy production, European Commission, 2024, https://joint-research-centre.ec.europa.eu/jrc-news-and-updates/amazon-region-2022-and-2023-deforestation-forest-degradation-and-risk-growing-soy-production-2024-02-28_en
20 Joint Research Centre, The Amazon region in 2022 and 2023: Deforestation, forest degradation and the risk of growing soy production, European Commission, 2024, https://joint-research-centre.ec.europa.eu/jrc-news-and-updates/amazon-region-2022-and-2023-deforestation-forest-degradation-and-risk-growing-soy-production-2024-02-28_en
21 UN, What is happening to agrobiodiversity? FAO, 1999, https://www.fao.org/4/y5609e/y5609e02.htm
22 Lakhani, N., Chang, A. & Witherspoon, A., Our food system isn't ready for the climate crisis, *The Guardian*, 2022, https://www.theguardian.com/food/ng-interactive/2022/apr/14/climate-crisis-food-systems-not-ready-biodiversity
23 Davis, D. R., Epp, M. D. & Riordan, H. D., Changes in USDA food composition data for 43 garden crops, 1950 to 1999. *Journal of the American College of Nutrition, 23* (6), 669–682, 2004, DOI:10.1080/07315724.2004.10719409
24 Reda, F. M. M. A., et al., Impact of micronutrient deficiencies on child development. *Journal of Medical Genetics and Clinical Biology, 1* (6), 261–267, 2024, DOI:10.61796/jmgcb.v1i6.674
25 Eldridge, H. M., *The avocado debate* (1st ed.). Routledge, 2023, DOI:10.4324/9781003371915

26 BBC News, Pink lady v the British apple, *BBC*, 2013, https://www.bbc.com/news/magazine-24578762
27 Aggret, Sammy, et al., *Advances in poultry genetics and genomics*. Burleigh Dobbs Science Publishing, 2020, DOI:10.1201/9781003047735
28 Doublet, A., et al., The impact of genomic selection on genetic diversity and genetic gain in three French dairy cattle breeds. *Genetics Selection Evolution, 51* (1), 52, 2019, DOI:10.1186/s12711-019-0495-1
29 GOV.UK, Genetic Technology (Precision Breeding Act 2023), Acts of Parliament, 2023, https://www.legislation.gov.uk/ukpga/2023/6/pdfs/ukpga_20230006_en.pdf
30 GOV.UK, Genetic Technology (Precision Breeding Act 2023), Acts of Parliament, 2023, https://www.legislation.gov.uk/ukpga/2023/6/pdfs/ukpga_20230006_en.pdf
31 Coe, S., et al., The effect of the war in Ukraine on UK farming and food production, House of Commons Library, 2022, https://researchbriefings.files.parliament.uk/documents/CDP-2022-0147/CDP-2022-0147.pdf
32 GOV.UK. Chapter 3: Farming income. In: *Agriculture in the United Kingdom 2024 – Annual statistics about agriculture in the United Kingdom*. DEFRA, 2025, https://www.gov.uk/government/statistics/agriculture-in-the-united-kingdom-2024/chapter-3-farming-income#summary
33 Sellgren, K., Families 'can't afford to follow healthy diet guidance', *BBC News*, 2018, https://www.bbc.com/news/education-45420295
34 Brinsden, H., Eating away at productivity: The toll of diet-related ill health, The Food Foundation, 2024, https://foodfoundation.org.uk/sites/default/files/2024-09/Eating%20away%20at%20productivity.pdf
35 Campbell, D., UK's unhealthy food habits cost £268bn a year, report finds. *The Guardian*, 2024, https://www.theguardian.com/politics/2024/nov/15/uk-unhealthy-food-costs-268bn-a-year-report-food-farming-countryside-nhs
36 Stiebahl, S., Research briefing: Obesity statistics, House of Commons Library, 2025, https://commonslibrary.parliament.uk/research-briefings/sn03336/
37 UN, Review of Maritime Transport, UNCTAD, 2024, https://unctad.org/publication/review-maritime-transport-2024
38 GOV.UK, Competition, choice and rising prices in groceries, Competition & Markets Authority, 2024, https://assets.publishing.service.gov.uk/media/64b80adaef5371000d7aeefb/Competition__choice_and_rising_prices_in_groceries.pdf
39 ETC Group, Agrochemicals and commercial seeds. In: *Food Barons 2022. Crisis profiteering, digitalization and shifting power*. ETC Group, 2022, https://www.etcgroup.org/files/files/food-barons-2022-full_sectors-final_16_sept.pdf
40 GRAIN & The Institute for Agriculture and Trade Policy, The Fertiliser Trap, The rising cost of farming's addiction to chemical fertilisers, IAPT, 2022, https://www.iatp.org/the-fertiliser-trap
41 Vogler, Penn, *Stuffed: A history of good food and hard times in Britain*. Atlantic Books, 2023.
42 UN, World Food and Agriculture – Statistical Yearbook 2022, FAO, 2022, DOI:10.4060/cc2211en
43 GOV.UK, UK Food Supply Sources. In: *United Kingdom Food Security Report 2021*. DEFRA, 2022, https://www.gov.uk/government/statistics/united-kingdom-food-security-report-2021/united-kingdom-food-security-report-2021-theme-2-uk-food-supply-sources
44 Lang, T., *Feeding Britain: Our food problems and how to fix them*. Pelican, 2020.
45 The Economist, Ukraine's front-line farms battle Russians and weather, 2025, https://www.economist.com/europe/2025/07/13/ukraines-front-line-farms-battle-russians-and-weather

46 Welsh, C. & Glauber, J., Food as the "silent weapon": Russia's gains and Ukraine's losses, Centre for Strategic and International Studies, 2024, https://www.csis.org/analysis/food-silent-weapon-russias-gains-and-ukraines-losses
47 Brookings Office of Communications, Analysing Israel's strategy in Gaza (online event), 2024, https://www.brookings.edu/events/analyzing-israels-strategy-in-gaza/
48 Feingold, S. & Botwright, K., Tracking tariffs: Key moments in the US-China trade dispute, World Economic Forum, 2025, https://www.weforum.org/stories/2025/06/trumps-us-china-trade-tariffs-timeline/

4 How to build food resilience

While progress at the farm and retail levels is vital, the most transformative change in the food system depends on government intervention. Farmers, processors, and retailers operate within the policy and economic landscape that governments shape, and their ability to innovate or adapt is either enabled or constrained by those conditions. Subsidies, trade policies, environmental regulations, and public procurement standards all send powerful signals that influence what is produced, how it is produced, and who benefits. Without coherent policy leadership, even the most committed efforts by individual actors risk being undermined by conflicting incentives or market pressures. Both national and local governments must therefore create the enabling environment that makes positive shifts possible—through strategic regulation, targeted investment, and coordination across departments such as agriculture, health, education, and environment. This includes aligning fiscal and trade policies with goals for climate resilience, nutrition, and equity. Governments also have a critical role in fostering transparency, supporting research, and ensuring that smallholders and marginalised communities are not left behind. Ultimately, policy leadership is essential to embed resilience, fairness, and sustainability throughout the food system—aligning public goals with the everyday realities of farmers, retailers, and consumers alike, and steering collective action towards long-term transformation.

Achieving long-term food stability in the UK requires coordinated policy action across every sub-category of the food system, guided by both targeted interventions and an overarching, integrated vision. Each dimension of food resilience—spanning production, distribution, access, and environmental—demands tailored policy responses that address its specific challenges while remaining connected to the broader system. Because food resilience is inherently complex and deeply interdisciplinary, significant overlap exists between these categories. However, a clear policy focus within each area is essential to ensure coherent implementation and measurable outcomes. The following framework outlines the most urgent policy interventions needed to strengthen the UK's food resilience. It recognises that no single policy can operate in isolation; progress depends on the alignment of actions across sectors such

DOI: 10.4324/9781003581871-5

as agriculture, health, trade, and environment. Establishing distinct policy objectives within each sub-category helps clarify responsibilities, streamline delivery, and prevent duplication of effort. At the same time, maintaining an integrated national vision ensures that these efforts collectively contribute to a more resilient, equitable, and sustainable food system capable of withstanding economic, social, and environmental pressures in the years ahead.

Solutions for economic resilience

For farming to endure, it must be economically viable. Yet across the UK, small and medium-sized farms are faltering under mounting pressure from volatile markets, low supermarket prices, and rising production costs. Contracts are frequently altered or dropped, while input costs for energy, fertiliser, and feed have soared—exacerbated by global shocks. Policy uncertainty compounds these challenges. The transition from EU subsidies to the ELM has created confusion, while labour shortages and rising borrowing costs add further strain. These pressures have left many unable to reinvest, innovate, or plan ahead. Without stronger protections, fairer trading practices, and consistent long-term support, more farmers will exit the sector—weakening rural economies and undermining the UK's food security. Actions that would deliver economic resilience within the food system include:

Reassurance over ELM funding

To deliver genuine economic resilience within the food and farming sector, the Treasury must reverse the planned 2.7% annual funding cuts to Defra from 2025/26 to 2028/29 and instead commit to a long-term roadmap for how funding will be allocated through the ELM scheme.[1] This roadmap should clearly set out multi-year commitments and delivery milestones, providing farmers with the predictability they need to plan ahead with confidence. Currently, many farmers and growers operate in an atmosphere of profound uncertainty. Successive changes to ELM design, fluctuating timelines, and inconsistent information about the SFI have undermined confidence and delayed business decisions. As previously mentioned, the Defra Secretary of State has recently indicated an intention to reform the SFI. This redesign was announced on the heels of the decision to close the SFI in March 2025, when Government made the controversial decision to abruptly close the SFI scheme within the agreed 6-week notice due to the exhaustion of its £1.05 billion budget. The purpose of this review, according to Government, is to enhance the SFI's effectiveness and ensure equitable distribution of funds. While the announcement of the SFI's closure and redesign was poorly executed, the aspiration is a good one. The SFI today is not delivering to the best of its ability. The redesign creates a key opportunity for food sovereignty and agroecology to become the primary focus of government farming support in England. Farmers cannot build sustainable businesses—let alone invest in nature restoration or carbon-smart practices—without knowing what support they will receive or how long it will last. Agriculture operates on multi-year, often multi-decade, cycles. Decisions made today—whether to plant an orchard, invest in livestock infrastructure, or restore degraded soil—will determine farm viability for decades. Without policy stability, these forward-looking investments are impossible. Providing clarity, consistency, and long-term funding assurance through a well-communicated and stable ELM scheme would empower farmers to innovate, improve productivity, and adopt regenerative practices that take time to mature. A dependable policy environment is therefore the cornerstone of economic resilience—one that enables farming businesses to thrive while delivering public goods for climate, nature, and rural communities.

Combined authorities to invest in agricultural training and skill development

Many farmers currently lack access to the skills and training needed to transition towards more resilient and sustainable business models. As most are self-employed, they do not have access to the kind of structured continuing professional development that other sectors benefit from. Government should therefore play a proactive role in supporting lifelong learning within the farming community—helping farmers to strengthen both their business and technical capabilities. This support should be flexible and practical, reflecting the realities of farm life. It could include training on essential business tools, such as digital accountancy systems or website design, to help farmers diversify income streams and build more robust business models. Investment should also prioritise continuous professional development through peer-to-peer learning—such as mentoring schemes, cooperative knowledge-sharing networks, and farm clusters—that allow farmers to exchange expertise, trial innovation, and adopt new technologies. Investing in farmer education in this way would not only improve individual resilience but also enhance the overall stability and competitiveness of the UK's food system. Furthermore, farmers now have the ability to access many new technologies that can help improve the efficiency and sustainability of their production. However, they also need to be trained in how to use them and learn about new innovations that might benefit their business. Courses should cover emerging technologies and new sustainable farming techniques that can boost productivity while reducing environmental impact. To deliver this, national government should work in partnership with Combined and Local Authorities to establish a comprehensive skills pipeline that spans from entry-level training to higher-level vocational qualifications. This could include expanding Agriculture and Horticulture-focused Skills Bootcamps and integrating them into broader green skills strategies. To strengthen the economic resilience of UK farmers and food SMEs, national government should make targeted investment in skills, training, and business development a central pillar of its food and farming strategy. While combined authorities already hold devolved budgets for skills and employment, these funds are rarely directed towards the specific needs of agriculture, horticulture, or agri-food businesses. Redirecting a portion of these resources to support skills development in farming and food production would help build a new generation of producers and entrepreneurs equipped to manage the challenges of climate change, supply chain disruption, and shifting markets.

These continuing education courses should be complemented by new apprenticeships in nature-friendly and climate-resilient farming, alongside dedicated support for SMEs within the food supply chain to strengthen business planning, digital capacity, and market diversification. The UK currently faces a severe shortage of training opportunities focused on sustainable, commercially viable farming—particularly those grounded in agroecological and low-input systems. This leaves many new entrants without accessible routes into the sector, while an ageing farmer population threatens long-term food security and productivity. Importantly, programmes should focus on inclusion, helping people from underrepresented and marginalised backgrounds access careers in food and farming. By aligning national policy, devolved budgets, and local delivery, government can create a skilled, diverse, and forward-looking agricultural workforce that drives economic growth, enhances environmental resilience, and secures the UK's food future.

Ensure an inclusive international trade policy

All these interventions to secure an economically resilient future for UK farmers will be irrelevant if they are undercut by cheaper imports that have been produced to a lower environmental or social standard. Given that we are a small nation and exist in a globalised market, we will always need to import food. However, the Government has an obligation to ensure an inclusive and resilient trade approach that safeguards UK farmers. Future Free Trade Agreements (FTAs) must guarantee that imported products meet equivalent social, environmental, and animal welfare standards to those required of UK producers. The current imbalance—where UK farmers face higher production costs due to stricter environmental and welfare requirements—creates vulnerability to being undercut by cheap imports produced under weaker regulations. A future-proof trade framework should therefore embed non-regression clauses that prevent trading partners from lowering standards over time, alongside mirror clauses requiring equivalence on key production criteria such as pesticide use, antibiotic regulation, carbon emissions, and labour rights. To operationalise this, Defra and the Department for Business and Trade should work jointly to develop a Fair and Sustainable Trade Framework that aligns with the UK's Paris Agreement obligations and the UN Sustainable Development Goals. This framework should include robust participatory impact assessments—covering environmental, social, and human rights dimensions—before, during, and after trade negotiations. Future FTAs should also exclude Investor-State Dispute Settlements (ISDS) mechanisms, which risk prioritising corporate interests over public policy and environmental protection. Trade transparency must also improve. The Government should establish a statutory Trade and Standards Commission—with representation from farmers, environmental groups, and the devolved administrations—to scrutinise deals and ensure devolved concerns are respected. Furthermore, supply chain traceability and due diligence requirements should be strengthened to hold companies accountable for imported goods, ensuring they are free from deforestation, forced labour, and environmental harm. By embedding fair standards, transparency, and farmer inclusion into the trade policy process, the UK can foster equitable competition, uphold consumer trust, and ensure that trade supports—not undermines—the long-term sustainability and resilience of British farming.

Ensure affordable access to high-quality food through public procurement

Public procurement contracts hold significant potential to strengthen the economic resilience of small-scale producers when designed with local SMEs in mind. As major purchasers of food for schools, hospitals, and other public institutions, governments can use their buying power to create stable and predictable markets for local producers. By prioritising regionally sourced ingredients, procurement policies can help small producers secure reliable demand, reduce market volatility, and retain more value within local economies. However, this impact depends heavily on contract design. Procurement frameworks often favour large suppliers due to complex tendering processes, high compliance costs, and rigid volume requirements. Adjusting these mechanisms—such as breaking large contracts into smaller lots, simplifying application procedures, and incorporating social and environmental value criteria—would allow smaller enterprises to compete more effectively. Additionally, longer-term contracts can provide security for producers to invest in infrastructure, innovation, and workforce development. Well-designed public procurement policies also enhance the economic resilience of the broader food system by making healthy, locally sourced food more accessible to low-income and at-risk groups—ensuring that those who would benefit most from nutritious diets are directly supported through public spending. While much of the focus on public food rightly centres on school meals, this ambition must extend to all areas of public procurement—including hospitals, care facilities, prisons, and government canteens. Access to nutritious, sustainably produced food is essential for health and well-being across all public institutions. For patients and residents, better food supports recovery and quality of life; for those in the justice system, it improves dignity and stability.

Government should provide clear and practical guidance on how to meet the commitment for 50% of public sector food to be local or sustainable, ensuring this ambition delivers genuine progress towards the National Food Strategy's goals of improving health, reducing emissions, and supporting British producers.[2] To achieve this, "locally produced" food should be defined as products that are both grown or made in the UK and sourced from small- or medium-sized enterprises or bought through direct-to-farmer supply chains, such as dynamic procurement platforms. While we will discuss the ways in which government can deliver environmental resilience within food and farming, it is worthwhile highlighting, while we're here, that "certified to higher environmental standards" should mean that the food is certified organic produce. Of the 50% target, at least 20% should consist of UK-sourced certified organic fruit, vegetables, or pulses, measured by weight rather than monetary value to ensure meaningful environmental impact. To embed these standards consistently, the Government should expand and strengthen the Government Buying Standards (GBS) across all public institutions, ensuring alignment with national climate and nature recovery targets. Enforcement should be backed by mandatory reporting through Ofsted and other relevant inspectorates, guaranteeing transparency and accountability. This would ensure that public money is invested in food that nourishes people, sustains farmers, and protects the environment.

Wales and Scotland are already leading the way on public procurement, demonstrating how devolved governments can use their purchasing power to strengthen domestic farming, support rural economies, and improve public health. Both nations have taken significant steps to prioritise local and sustainable food sourcing across schools and public institutions—showing what a joined-up approach to food, farming, and health can achieve. Yet, both could go further to maximise these benefits and embed agroecological and organic production at the heart of their food strategies. In Wales, the Government could build on its existing commitments by requiring that 25% of all primary school vegetables are locally grown and organic, achievable by spending just an additional 3p per meal. This modest investment would deliver high returns for health, local economies, and farm viability. In Scotland, a stronger emphasis on sustainable sourcing could be achieved by committing to 15% of food in public contracts being organic by 2030, with clear interim milestones to drive progress. This target would stimulate demand for local organic production, reduce climate impacts, and raise the quality of public meals. To make this feasible, the Scottish Procurement Directorate should update and relaunch "Catering for Change", embedding sustainability, health, and biodiversity goals alongside practical case studies for local authorities, schools, and hospitals. Food Standards Scotland should complement this by publishing new guidance on legumes and allergens to encourage wider use of beans, peas, and lentils as affordable, low-carbon proteins.

Local and regional authorities have a vital role to play in enabling more people to access agroecological and organic fruit and vegetables, ensuring that public food spending delivers social, environmental, and economic benefits within communities. A first step should be to appoint catering support roles, such as Food Improvement Officers, who can work directly with kitchen teams to develop recipes and menus that make the most of seasonal, locally grown produce. Practical workshops and peer learning sessions can help caterers adapt confidently to using fresh ingredients from smaller suppliers. To complement this, councils should establish supply chain facilitators who act as connectors between small growers, SMEs, and public sector buyers. These roles can help producers navigate procurement processes, build local partnerships, and ensure that smaller suppliers can meet the logistical and compliance demands of large contracts. Local authorities can also strengthen their commitment by adopting local food purchasing policies aligned with frameworks such as the London Food Purchasing Commitment, which prioritise sourcing from sustainable and community-based producers.[3] This commitment required London councils to report the total ingredient budget spent on food that supports fairer, more equitable, sustainable and farmer-focused food production, aimed to reduce food waste by 50% and per plate emissions by 38% by 2030. Using London's example, procurement processes should be redesigned to be more SME-friendly, including flexible payment schedules and smaller, more accessible contract lots that level the playing field for independent producers. Finally, tenders should include clear sustainability and locality weightings to reward suppliers with strong environmental credentials—such as organic or agroecological certification—and to ensure that public procurement actively supports local economies and community well-being.

Solutions for supply chain resilience

Building a more resilient UK food supply chain will depend on bold and targeted policy change. Today, as we have discussed, only four retailers control around 70% of the grocery market, creating bottlenecks as restrictive as global shipping choke points like the Suez or Panama Canals.[4] Policy must therefore focus on rebalancing the food economy—reducing dependence on both international trade corridors and domestic retail oligopolies. By investing in the rollout of new technologies to smooth existing supply chains while encouraging new entrants into the food system to break the current supermarket oligopoly, government can foster a system with multiple, overlapping routes to market. True resilience will come not from scale or efficiency alone, but from diversity, flexibility, and connection across the entire food chain. Actions that would deliver supply chain resilience within the food system include:

Embrace new supply chain technologies

Logistics technologies have the potential to transform the resilience of the UK's food supply chain by reducing vulnerability to shocks and chronic pressures while maintaining consistent access to food. Emerging technologies can help create more adaptive and efficient systems capable of responding to disruption with agility. Digital tools such as AI-driven logistics, Internet of Things sensors, and blockchain traceability can provide real-time data across the supply chain—allowing producers, distributors, and retailers to anticipate delays, minimise losses, and reroute supplies when crises occur. Smart storage, new packaging innovations and cold-chain technologies can reduce spoilage during extreme weather or transport disruptions, ensuring food reaches consumers safely and efficiently. To unlock this potential, government policy must play an enabling role. Targeted investment through innovation funds and public–private partnerships could accelerate technologies with proven benefits for food system resilience. One clear opportunity is to leverage and adapt the UK Research and Innovation (UKRI) Transforming Food Production (TFP) Challenge Fund, which already supports agritech innovation but could be expanded to include supply chain technologies and packaging and transport innovations.

Create a national land-matching scheme

Land-matching schemes present a vital opportunity to revitalise British agriculture, ensuring continuity between generations while opening pathways for a more inclusive and diverse farming sector. In France, successful initiatives such as Réseau National des Espaces-Test Agricoles (RENETA) and Terres de Liens have demonstrated how structured land matching can ease succession and strengthen rural resilience.[5] These programmes connect retiring farmers with new entrants and provide transitional "test farms" where prospective farmers can gain experience, trial enterprises, and learn directly from established producers before assuming full management. This mentorship-based approach preserves invaluable local knowledge while giving new farmers the time, security, and confidence to build viable businesses. Adapting similar models in the UK could address the twin crises of an ageing farming population and the steep barriers to entry faced by new growers—especially those from underrepresented groups, including people of colour, women, and first-generation farmers. Traditional routes to land ownership or tenancy are often inaccessible due to high costs and entrenched networks. A national land-matching programme, backed by government funding and delivered in partnership with local authorities and farming organisations, could help bridge this gap. By pairing experienced farmers with motivated new entrants, the UK could facilitate intergenerational handovers that maintain agricultural productivity, enhance biodiversity through new agroecological approaches, and diversify the people who steward the land. This is not just succession planning—it is a step towards a fairer, more resilient food future.

Greater local start-up support for agricultural SMEs

To strengthen regional food resilience and unlock local economic potential, local and combined authorities should establish Agricultural Start-Up Funds to support small- and medium-sized food and farming enterprises. SMEs are central to building a fairer, greener food economy—creating jobs, supporting innovation, and keeping more money circulating locally. Every £10 spent in a local food business generates nearly £25 for the local economy, demonstrating the high return on investment such initiatives can deliver for local authorities. A Start-Up Fund would provide start-up seed funding, low-interest loans, and targeted capital grants for essential infrastructure such as polytunnels, cold storage, and irrigation systems. It should also include integrated business and sustainability advice to help enterprises plan for long-term growth. Support should be tailored to urban and peri-urban horticulture, where land and infrastructure pressures are particularly acute. Local and combined authorities can play a pivotal role in enabling local food businesses—including peri-urban farms and market gardens—to catalyse a thriving local food sector that drives jobs, resilience, and inclusive economic growth. A model for such investment could be the Agri-Tech and Food Technology in Eastern England programme that provides up to £4.5 million in grant funding to support new innovators across the region's food and farming sectors.[6] Delivered through Innovate UK and local authorities, the fund strengthens regional economic resilience and positions Eastern England as a leader in sustainable agri-food innovation and technology development.

To support new entrants into farming, local councils should also re-establish our county farms as incubators. County farms—also known as council farms—originated from the Smallholdings and Allotments Acts (1908–48), which empowered county councils to buy land and lease it to aspiring farmers.[7] The goal was to create a ladder of opportunity for people without inherited land, allowing them to gain practical experience, build capital, and progress towards independent ownership or larger tenancies. By the mid-20th century, local authorities managed more than 400,000 acres across England and Wales, offering thousands of entry-level holdings. However, decades of asset sales, funding cuts, and political shifts have eroded this network. Today, fewer than 200,000 acres remain, concentrated in a handful of counties such as Cambridgeshire, Norfolk, and Lincolnshire.[8] Many estates now have limited turnover and longer leases, restricting new entrants' access. Despite this decline, surviving county farms continue to demonstrate the value of structured tenancies combined with mentoring and business training. They remain one of the few public tools for supporting land access, rural skills, and generational renewal in farming. Reviving the county farms system would help support supply chain resilience by ensuring new entrants can come through, more start-ups succeed. For county farms to reclaim their full potential requires protecting the land base, expanding access, and embedding structured support for tenants. First, the Government should safeguard remaining estates through a Public Farmland Covenant, preventing local authorities from selling off this vital public asset. Stable ownership would give councils the confidence to plan long-term for training and succession rather than short-term revenue. Alongside protection, councils should be funded and incentivised to expand their holdings—acquiring or leasing land using grants from Defra's New Entrants Support Scheme—to open more opportunities for first-time farmers.[9] Yet access alone is not enough: success depends on mentorship and skills development. Each new tenancy should come with tailored business planning, technical guidance, and peer learning, delivered through agricultural colleges and the Agriculture and Horticulture Development Board. Together, these measures would transform county farms into modern incubators—secure, well-supported spaces where new entrants can learn, innovate, and grow towards independent farming careers.

Solutions for market resilience

Tackling the lack of resilience within agricultural markets will require measures that rebalance power across the food system and create space for fairer, more diverse participation. Addressing market concentration is central to this goal. Stronger competition policy and tighter scrutiny of mergers are needed to prevent further consolidation and to ensure that farmers and small food businesses can compete on fair terms. Government can also promote resilience by supporting new local retail businesses to set up—reducing dependence on a few global corporations. Policies that encourage cooperative models and farmer-led enterprises can help producers gain bargaining power and retain more value from what they grow. Ultimately, building resilience means moving beyond a system dominated by a handful of global actors towards one rooted in diversity—of ownership, scale, and geography. By empowering smaller players and decentralising control, governments can help rebuild a food system that is fairer, more adaptable, and better able to withstand shocks. Actions that would deliver market resilience within the food system include:

Strengthen the Grocery Code Adjudicator

Improving supply chain fairness through reform of the grocery sector is essential to strengthening the economic stability of UK farmers and would directly advance the 2025 National Food Strategy's objective of creating "fairer, more transparent supply chains" and help ensure that farmers receive a fair return for their work.[10] At present, farmers across the UK face intense and unsustainable price pressures from powerful retailers and intermediaries. The current regulatory framework leaves significant gaps in oversight, particularly around pricing practices, contract cancellations, and late payments. The Grocery Code Adjudicator (GCA) has already been bold in their investigation into Amazon's grocery practices, but their ambition needs to be supported to go further. A stronger, better-resourced GCA, and also the associated Agricultural Supply Chain Adjudicator (ASCA), would not only protect farmers from exploitative practices but also foster a culture of fairness and accountability across the entire food chain. By embedding these reforms, the UK could build a food economy that rewards sustainability, supports domestic production, and ensures that resilience and profitability go hand in hand. This could be further supported by mandatory and honest front-of-pack product labelling for human health and animal welfare.

Spain's Food Chain Law, strengthened in 2021, offers an ambitious model for promoting fairness and transparency in the agri-food sector from which the UK would do well to draw inspiration.[11] The law requires all food-supply contracts to be written, transparent, and based on prices that at least cover verified production costs—explicitly banning the practice of buying or selling below cost. It also sets out a "black list" of unfair trading practices, such as unilateral contract changes or late payments, and empowers Spain's Food Information and Control Agency (AICA) to enforce compliance through inspections, fines, and a new digital register of contracts. By extending protection across the entire supply chain, not only between farmers and retailers but also processors, distributors, and even some international transactions, the Spanish model aims to rebalance power and improve the stability of farm incomes. For the UK, similar reforms could address long-standing imbalances between producers and large buyers. Strengthening and expanding the remit of the GCA and ASCA, introducing mandatory written contracts across sectors, and piloting cost-of-production safeguards would build on existing fair-dealing rules and ensure that shocks in energy or input prices do not simply cascade down to farmers. Spain's example demonstrates that clear rules, transparency, and strong enforcement can make agricultural markets both fairer and more resilient.

Improving competition through anti-trust policies

A more competitive, transparent, and resilient agricultural inputs industry is vital to the UK's long-term resilience of the agricultural sector. Lessons from both Senator Chuck Grassley's efforts in the USA and the Biden administration's wider anti-trust agenda point to a clear principle: concentrated input markets—whether in fertiliser, seeds, pesticides, or digital technology—distort prices, suppress innovation, and weaken the bargaining power of farmers. The UK can act decisively to prevent similar structural risks by embedding competition and openness into its regulatory framework. First, the Government should establish a specialised Agricultural Inputs Market Authority (AIMA) within the Competition and Markets Authority (CMA), co-staffed with Defra economists and technical experts. This unit would monitor market concentration and coordinate competition enforcement across fertiliser, seed, pesticide, machinery, and ag-tech sectors. Modelled on the Biden-era USDA–DOJ partnership, the AIMA could publish periodic market studies and concentration reports, creating continuous oversight rather than reacting only when mergers arise. A top priority for this body should be a Fertiliser Competition and Resilience Inquiry, assessing pricing transparency, supply vulnerabilities, and the dominance of a few multinational suppliers. The inquiry would underpin a new Fertiliser Innovation and Competition Fund, using targeted grants and guarantees to support domestic production of low-carbon and bio-based fertilisers, diversify import routes, and stimulate SME entry. By coupling competition analysis with resilience funding, the UK can align its food and energy security agendas.

Reforming control over agricultural technology and machinery is equally critical. A right-to-repair and data-portability regime under the Digital Markets, Competition and Consumers Act (2024) should compel large equipment manufacturers and digital platforms to share diagnostic tools, repair manuals, and farm data with users and independent technicians. This will reduce lock-in, extend equipment life, and open opportunities for software and service innovation by smaller firms. Complementing these structural reforms, the CMA could publish clear guidance on anti-competitive data sharing, targeting third-party benchmarking and information exchanges that enable suppliers to coordinate prices or output. This would mirror US action against data-based collusion and ensure the UK's digital agricultural transition supports, rather than restricts, competition. This approach of greater data sharing and transparency should extend to competition and diversity metrics being built into Defra's regular Food Security Reports. This would help ensure market concentration is treated as a systemic risk alongside climate, trade, and labour. By linking competition policy to food system resilience, the UK can foster an inputs sector that is innovative, fair, and strategically secure.

Local authorities to create new opportunities for local retail

However, more than simply penalising or halting the worst and most unfair practices in the supply chain, we need to proactively build opportunity and resilience into the retail system. This means moving beyond compliance and focusing on creating a more diverse network of routes to market that reduce dependency on dominant supermarket supply chains. To achieve this, government and industry must work together to challenge the excessive control held by large retailers, support the growth of alternative and cooperative distribution models, and invest in infrastructure that connects small- and mid-sized producers directly with consumers, public institutions, and local businesses—ensuring a fairer, more resilient food economy. Recent reforms to England's planning framework offer an important opportunity to create more community food spaces and alternative retail opportunities that strengthen local economies and social resilience. Updates to the National Planning Policy Framework (NPPF) give local authorities greater flexibility to embed social and economic value within spatial strategies. In parallel, the Planning and Infrastructure Bill (2025)—could further empower councils to reimagine high streets and public spaces by expanding delegated powers and simplifying local decision-making. Together, these changes open the door to a more community-oriented planning agenda—one that recognises that access to food, social connection, and local enterprise are essential components of sustainable urban development. Local authorities should seize this opportunity by identifying and repurposing underused land, empty retail premises, and disused community buildings for food-related social enterprise. This could include cooperative grocers, community cafés, local markets, and shared kitchens—spaces that support independent businesses, create jobs, and provide residents with affordable, healthy options beyond supermarkets. Planning strategies should actively embed these spaces into new residential and regeneration projects, ensuring that developments include accessible areas for community food use, social dining, and local retail. To make this vision achievable, national and local governments should align planning policy with social and economic goals by using mechanisms such as business rate relief, Section 106 agreements, and community asset transfers to prioritise food-based community projects. High streets, too, can be revitalised through mixed-use planning that integrates small-scale retail and food enterprises into public spaces. By treating food access and community enterprise as critical infrastructure, planning reform can deliver healthier, fairer, and more connected neighbourhoods—where good food, social interaction, and local business sit at the heart of community life.

Beyond repurposing vacant spaces to create new retail opportunities, government could push to create alternative routes to market that would allow more resilience. Across the UK, the closure of many wholesale markets has left small- and medium-sized food businesses with limited routes to distribute their products outside of supermarket supply chains. To address this growing gap and build food system resilience, each Combined Authority should establish a regional food distribution hub dedicated to supporting SMEs and independent producers. These hubs would act as modern, multi-functional spaces for the collection, storage, and distribution of local produce—replacing the vital role once played by wholesale markets. An example of a recent development that highlighted the importance of wholesale markets is that of the Victorian Fruit and Vegetable Market in Dublin. After being closed for 6 years, officials in Dublin City Council recently realised its importance from a commercial and cultural heritage perspective, and scheduled its reopening for 2027. Chief Executive Richard Shakespeare said that "the refurbishment will provide a high-quality, sustainable retail food market, enhancing both the local economy and the public realm", becoming a hub for community and local businesses.[12] Each regional hub would also serve as an incubator for innovation, offering shared logistics, cold storage, and digital infrastructure that allow local enterprises to grow, collaborate, and access new markets. Combined Authorities should integrate food infrastructure planning into their Spatial Development Strategies and allocate capital funding for the creation and maintenance of these hubs, potentially through existing levelling-up or regional growth funds. By investing in regional food distribution, government can help revitalise local economies, strengthen SME competitiveness, reduce dependency on supermarket monopolies, and create a fairer, more sustainable food system for the UK. Ultimately, these hubs could form a coordinated national network to improve supply chain efficiency and regional food resilience.

Solutions for climate resilience

A sustainable future for food depends on confronting the climate crisis directly, and actively focusing on resilience in farming and food production. Climate resilience in agriculture means the ability of food systems to adapt to and mitigate climate impacts through methods like crop diversification, efficient water use, and biosphere-smart innovations. It also requires environmental resilience—the capacity of ecosystems to withstand shocks such as drought, pests, and extreme weather while maintaining vital functions like soil health, water cycling, and biodiversity. Strengthening both forms of resilience is essential to ensure long-term food stability in the face of increasing climate-related challenges. This external vulnerability highlights the need for a more resilient and sustainable UK food system. Diversifying supply sources, supporting sustainable farming at home and abroad, and reducing food waste are all critical steps. Strengthening local food production and improving ecosystem health can also help buffer the UK against international environmental shocks. Actions that would deliver climate resilience within the food system include:

Adopt a land use framework based on land sharing

The Government has increasingly been talking about a "land sparing" approach that would separate areas of intensive agricultural production from areas set aside for nature. The aim of the land sparing approach is to meet food demand on smaller, high-yielding farmland while "sparing" other land for conservation. It contrasts with "land sharing", which integrates biodiversity within farmed landscapes. The "land sharing" approach maintains that all farming should have a lighter environmental footprint so that farming functions in harmony with nature. The land sparing model is gaining popularity as a solution for nations that are facing limited agricultural land and development pressures. On paper, the model appears to reconcile food production with conservation. In practice, however, it can trade off food production and nature objectives against one another. Intensification, even when marketed as "sustainable", often increases fertiliser use, irrigation demands, and mechanisation, all of which can raise emissions, degrade soils, and reduce on-farm biodiversity. Moreover, "land sparing" often underestimates the emissions from the "non-spared" land and the broader climate impacts of intensification itself. The intensive production that underpins the land sparing approach is incredibly damaging to the environment since diffuse pollution does not stay in one place but impacts the wider ecosystem.

No more clearly this is seen in the rise of intensive industrial livestock units. No amount of rewilded land can offset the damage that they cause to human and animal well-being and planetary health. These facilities place heavy pressure on local environments—degrading soil and water, increasing ammonia emissions, and undermining biodiversity. To this end, at a minimum, government should introduce significantly tighter planning controls for large-scale intensive livestock operations, particularly indoor poultry units. Planning authorities should therefore require full environmental impact and cumulative pollution assessments before any application is considered, alongside clear evidence that the proposed site will not harm local ecosystems, watercourses, or air quality. Planning approval should only be granted where developments can show a net environmental benefit and long-term resilience to climate impacts.

In some countries, like the US, land sparing approaches might have more success. They have huge expanses of open land that can be protected as national parks, and thousands of hectares of land can be given over to nature in perpetuity. Consequently, it has a better shot of working. But in small, densely populated countries, like the UK, these pressures are compounded by competing demands for land—housing, infrastructure, renewable energy, and recreation—alongside food production and habitat conservation. Focusing the land use framework on land sparing would be misguided: it risks marginalising smaller, mixed, and low-input farms that support biodiversity, rural livelihoods, and cultural landscapes to the benefit of larger industrial producers. A land sparing model could deepen regional inequalities and overlook multi-functional land uses—such as agroecology and native breed grazing—that deliver food, carbon, and nature together. To embed true environmental resilience into farming, government will need to embrace a land-sharing vision and adopt a holistic, integrated perspective. Land use policy cannot afford to treat climate change, biodiversity, and human health as separate concerns; these issues are deeply interconnected, and action in one domain inevitably affects the others. Policymakers should therefore seek to maximise co-benefits—such as improving soil health, reducing emissions, and enhancing food security—while actively working to minimise trade-offs.

Yet, it is not just agricultural land that we need to share with nature. We need to consider land sharing principles for urban and peri-urban areas, too. The UK needs to ensure that new housing and mixed-use developments are designed with food and health at their core. Access to healthy, affordable food should be seen as a fundamental part of community infrastructure, not an afterthought. Where a person lives profoundly shapes their relationship with food. Those living in so-called *food deserts*—areas with limited access to affordable, nutritious food—or *food swamps*, where unhealthy options dominate, face significant barriers to well-being. This problem often arises when developers fail to consider food access and green space in their plans. Across the country, too many communities lack proximity to fresh produce outlets or green, productive spaces that connect people with the land that sustains them. Many new developments still prioritise density over liveability, resulting in areas with little to no access to high-quality green space. Such spaces are more than just amenities—they are vital for mental and physical health, providing opportunities for recreation, relaxation, and social connection. Yet, access to them remains deeply unequal. Disadvantaged communities often face poorer air quality, less green infrastructure, and fewer safe, inviting outdoor areas, exacerbating health inequalities and social isolation. Equitable access to green space must therefore be seen as an issue of social and environmental justice, essential for improving national health and productivity. To address these challenges, the Government should ensure that food and green infrastructure are integrated into urban planning policy and national development standards. Local and combined authorities can play a pivotal role by requiring developers to allocate space for community food growing, urban orchards, or rooftop gardens; incorporating shared cooking, dining, and retail spaces into housing design; and embedding food access considerations into health impact assessments. Public Health should be a key partner in planning decisions to ensure developments actively promote healthy eating and community well-being. Meanwhile, use of schemes on vacant or underutilised land should also be encouraged for food growing and community food enterprises, helping to reconnect people with food, foster social cohesion, and strengthen local resilience.

Better support for sustainable farming via ELM

As previously mentioned, Defra's commitment to redesign the SFI offers a vital and timely opportunity to embed environmental and climate resilience more deeply within the UK's food system. This review period represents far more than a technical adjustment—it is a chance to realign the entire ELM framework so that it actively rewards farming systems that regenerate natural capital, rather than simply mitigate harm. To achieve genuine systemic change, SFI payments must encourage sustainability across entire holdings rather than isolated parcels of land. Farmers who adopt regenerative practices throughout their operations should receive full recognition and financial support, while partial compliance should no longer qualify for public funding. This approach would prevent the continuation of "business as usual" intensive farming alongside token environmental measures, ensuring that public investment delivers public goods—healthy soils, clean water, thriving wildlife, and nutritious food. By shifting incentives towards whole-farm sustainability, Defra can help farmers move beyond piecemeal "greening" measures and instead adopt integrated agroecological and organic approaches that build soil health, restore biodiversity, and strengthen the sector's capacity to adapt to climate shocks. By using this window of reform to reorient incentives, Defra can turn the SFI and wider ELM programme into the backbone of a resilient, nature-positive food system—one where environmental health, economic viability, and public well-being are no longer competing objectives, but mutually reinforcing goals. In order to maximise the impact, Defra needs to provide a free or low-cost, high-quality advisory service to accompany all SFI recipients so that farmers receive expert guidance on which actions will deliver the most environmental benefit on their farms. Simply offering a menu of actions isn't enough: farmers need support to assess their land's ecology, understand what practices will restore soil, biodiversity and resilience, and receive advice tailored to their region and farm type. Without this "right place, right action" focus, there is a risk of funds being spent on low-impact measures that don't add up to meaningful change.

Ensuring that organic and agroecological farming are better supported within the scheme would help to deliver an environmentally sustainable future for UK farming. One transformative measure would be to set a national target for expanding organic production to 10% of total farmland. This would not only signal government ambition but also deliver tangible benefits for climate resilience and soil regeneration. Organic and agroecological systems are notably more environmentally sustainable and climate resilient than regenerative models. Regenerative farming has value as a first step in a farm becoming more environmentally sustainable and is often an important and significant first step. However, all farmers should be actively encouraged and supported to advance their environmental stewardship beyond those initial steps. The environmental benefit of organic goes far beyond that of regenerative. It improves soil structure and carbon sequestration through diverse rotations, grass–clover leys, and reduced input dependency, while enhancing water retention and biodiversity both above and below ground.[13] Organic production also reduces the damage caused by agrochemicals by farming without pesticides and using natural fertility, instead of synthetic fertilisers. Such systems also support farm profitability over the long term by reducing exposure to volatile input markets and climate risks. Creating an organic target would help ensure that there was momentum towards organic conversion and that conversion was supported. Embedding this target within the forthcoming

National Food Strategy Action Plan would help address its critique of the "invisibility of nature" in food policy and ensure that the market begins to reward the environmental services farmers provide. Adopting this target could also create the mandate for supportive policy interventions, such as an SFI support option to cover conversion and certification costs. Conversion costs can often deter farmers from switching over. Under UK organic standards, land needs to be managed along organic principles for 3 years before the produce from that land can be labelled and sold as organic.[14] Farmers and growers are expected to take the economic hit of farming in a low-input manner without the benefit of the value-added organic label. This 3-year window of more expensive production costs but no way to balance that with higher sale prices puts many farmers off, or makes it impossible to afford, even if they might want to switch to more environmentally sustainable methods. Similarly, certification can easily cost up to over £500 for a first-time application and conversion, followed by an annual fee of around £500, an expensive outlay for a small-scale farmer on tight margins.[15] Covering these administrative burdens would help encourage more farmers to switch over to organic and manage their land more sustainably.

Better-resourced regulatory agencies to ensure compliance

All the incentives in the world will not work without the enforcement of environmental regulations. The Water Services Regulation Authority (OFWAT) has been in the spotlight for allowing water companies to pollute our waterways without consequences because the agency has been underfunded and hollowed out. The same is true of the Environment Agency (EA). It has been underfunded and understaffed. For example, in 2022–23, EA farm inspections covered only around 4% of farms in England, and the agency dropped more than 10,000 water-quality tests due to staffing and budget cuts.[16] Polluters no longer fear getting caught since the rate of spot inspections is so low. In order to ensure that polluters are held to account and that our environmental protections are as effective as possible, the EA needs to be given further resources. Without this, even the best-designed incentive schemes (such as those under the SFI) risk being undermined by the absence of credible deterrence.

In order to ensure greater climate resilience within our food and farming sectors, the Government should expand the Food Standards Agency's (FSA) remit to allow greater consideration of how environmental concerns interact with food safety issues. To date, the environmental impacts of farming and the consumer side of food regulation have been too siloed in UK legislation. The FSA has the remit of food safety and considers the hygiene of food. However, its remit does not expand to the wider food chain, and the agency avoids stepping into the space of advocating for a food system that is healthy for people *and* the planet. As far as Whitehall is concerned, Defra is responsible for farming but (despite its name) does not focus on food at a nutritional level since that responsibility falls to the DHSC, which does not have spare capacity to focus on much beyond trying to save the long-term future of the NHS. This hodgepodge of responsibilities and skill sets means that there is no minister and no department that is responsible for the food system as a whole. We need an FSA that has the ability to look from field to fork, from seed to spoon, and is adequately resourced to do so. By doing so, we can create a holistic understanding of food that is environmentally and nutritionally healthy.

Utilise the innovative power of technology for agroecology

Is there a role for technology in a climate-resilient model of UK farming? Or is technology antithetical to environmentally sustainable farming? Done right, technological innovation presents a powerful opportunity for the farming sector—not to industrialise farming, but to make it more sustainable, efficient, and resilient. Agroecology and technology are often seen as opposites: one rooted in ecology and tradition, the other in data and disruption. But this is a false dichotomy. Innovation can be shaped to serve ecological goals rather than undermine them. Small-scale robots, precision sensors, and digital mapping tools can all help farmers use fewer inputs, waste less water, and better understand their land—enhancing productivity while protecting biodiversity, reducing climate impacts and restoring soil health. For example, precision irrigation and soil sensors enable water and nutrient use that mirrors the natural cycles agroecology depends on, while AI-driven pest detection can reduce the need for chemical control entirely. Technology is not inherently industrial—it is a tool. Its impact depends on how, and by whom, it is used. When innovation is locally relevant, accessible, and designed to empower farmers, rather than extract profit, it can strengthen climate-sensitive farming rather than erode it. Open-source software and public investment can ensure that data-driven systems work for farmers and ecosystems, not just corporations. By embracing appropriate technology within a framework of transparency, and inclusion, the farming sector can leverage innovation to scale environmental resilience without scaling industrialisation. The future of sustainable farming should not be about rejecting technology—but about reclaiming it for people, planet, and food sovereignty. That said, the high cost of essential equipment remains a major barrier for new and small-scale food producers across the UK, particularly as access to innovative technologies like robotic weeders, drones, or soil sensors becomes increasingly important for sustainable production.

However, the high upfront costs of these technologies often make them inaccessible to smaller businesses, leaving innovation concentrated among larger farms and agribusinesses. To strengthen the sustainability and efficiency of UK farming, government and combined authorities should focus on improving access to emerging agricultural technologies that can help farmers produce more with fewer inputs. Tools such as robotic weeders, drone mapping systems, soil sensors, and precision irrigation controllers can significantly reduce fertiliser and pesticide use, optimise water consumption, and cut labour costs—helping farmers operate more sustainably while maintaining profitability. National and local governments should work together to establish community tool and machinery libraries, giving growers shared access to essential equipment—from basic tools such as seeders and repair kits to advanced precision technologies. These shared resources would reduce start-up costs, encourage collaboration, and make modern, sustainable farming methods accessible to a wider range of producers. Beyond improving efficiency and reducing financial pressure, such initiatives would help strengthen local food economies, build community connections, and foster peer learning among farmers. Supporting these libraries through capital grants or rural development funds would promote innovation, increase resilience, and empower more people to participate in the UK's transition towards sustainable, local food production.

The cost of delivering this via local and combined authorities could be supported through targeted funding streams—such as expanding the Farming Equipment and Technology Fund and expanding it to include shared-use models where local producer cooperatives could pool resources and share new technologies. In addition, training and technical support should be built into such schemes to ensure farmers can effectively use and maintain the equipment, maximising both economic and environmental gains. By reducing barriers to technology access, government and local authorities can help small-scale farmers improve efficiency, lower input costs, and reduce their environmental footprint—creating a more productive, climate-resilient, and digitally inclusive farming sector. Complementary investment in local agritech hubs, demonstration farms, and digital training programmes—delivered through existing mechanisms such as the Farming Innovation Programme (co-funded by Defra and UKRI)—would build technical capacity and ensure small- and medium-sized enterprises are central to this transition.[17] Planning and infrastructure policy should also recognise agritech as a critical part of national resilience, with funding for local agritech hubs, demonstration farms, and training programmes that build digital and technical capacity across the sector. Together, these innovations and supportive policies can create a smarter, more circular, and diversified food system—one capable of withstanding economic, environmental, and geopolitical shocks while ensuring that nutritious, sustainable food remains accessible to all.

Solutions for genetic resilience

Increasing genetic diversity in crops is essential for building climate resilience and ensuring long-term food security. As the effects of climate change intensify—bringing more frequent droughts, floods, heatwaves, and shifting pest pressures—agricultural systems must become more adaptable. Genetic diversity provides the foundation for this adaptability. It enables breeders and farmers to develop crop varieties with traits that help plants thrive under challenging and changing conditions, such as drought tolerance, salinity resistance, heat resilience, or rapid maturation. Many of these traits are found in traditional landraces or wild relatives of modern crops—genetic resources that have often been overlooked in industrial agriculture. By integrating this diversity into breeding programmes, we can reduce reliance on chemical inputs like fertilisers and irrigation, lower greenhouse gas emissions, and maintain stable yields even under climate stress. A genetically diverse farm also acts as a natural insurance policy. It is less likely to suffer total crop failure from a single pest outbreak or extreme weather event, making food systems more robust and livelihoods more secure. In livestock, greater genetic variation improves disease resistance and adaptability to local climates. To build sustainable, climate-smart agriculture, we must invest in conserving and utilising crop genetic diversity—through diversified breeding, seed saving, intercropping, and support for smallholder and Indigenous farming knowledge. Actions that would deliver genetic resilience within the food system include:

National seed sovereignty strategy

Seed saving and in situ conservation—the practice of maintaining and reproducing plant genetic diversity within farming systems and natural ecosystems—are vital for the resilience of UK agriculture. Seeds are the foundation of the food system, carrying the genetic diversity needed to adapt crops to changing climates, emerging pests, and evolving soil conditions. Yet, global seed diversity is shrinking at an alarming rate: industrial agriculture's focus on uniform, high-yield varieties has displaced locally adapted seeds, reducing the genetic base on which future breeding and adaptation depend. When farmers save and exchange seeds, they preserve regionally adapted traits—such as drought tolerance or pest resistance—that cannot easily be replicated in laboratories or gene banks. These living seed systems evolve with their environment, ensuring that agriculture remains dynamic, locally relevant, and resilient. For the UK, supporting in situ conservation and community seed networks should be recognised as a strategic priority for food security and environmental sustainability. The Government should develop a National Seed Sovereignty Strategy that invests in on-farm and community-based seed saving, strengthens collaboration between research institutions and growers, and ensures public funding for open-source breeding programmes focused on climate-resilient and agroecological crops. This mirrors the asks of the Global Partnership for Plant Conservation (GPPC) latest strategy for 2020–30 that outlines actions for both state and non-state actors to contribute to global biodiversity goals. Among these is utilising a combination of in situ and ex situ conservation methodologies, while also harnessing the knowledge and expertise of local communities and Indigenous people.[18]

This National Seed Sovereignty Strategy should include the reform of the regulations around genetic diversity in agriculture. Current rules, inherited from the EU, require that any seed variety sold commercially must be distinct, uniform, and stable (DUS). While these criteria were originally designed to protect farmers and ensure quality, they now disadvantage genetically diverse crops such as landraces, open-pollinated varieties, and evolutionary populations—all essential for climate resilience and local adaptation. These varieties, by their nature, evolve and vary over time, which makes them ineligible for the National List a legal register for plant varieties for sale. As a result, farmers and small-scale breeders are often prevented from legally marketing or exchanging them, limiting innovation and adaptation within UK farming. With post-Brexit autonomy, the UK has an opportunity to reform its seed laws to align with modern challenges. Defra should establish a new legal category for heterogeneous material, enabling diverse and farmer-bred varieties to be marketed legally, as already permitted under the EU's 2022 Organic Heterogeneous Material Regulation. Simplified registration systems, lower fees for small producers, and regional seed adaptation schemes would encourage innovation and safeguard crop diversity. Legalising and supporting genetically diverse seeds would strengthen resilience against pests, disease, and climate shocks—securing the UK's food future while empowering farmers to innovate and steward biodiversity.

Beyond productive plants and their seeds, there is an urgent need to protect the genetic diversity of our livestock. Defra's Native Breeds at Risk (NBAR) list currently identifies UK livestock and equine breeds with small breeding populations, supporting their conservation through agri-environment payments. From 2026, it will transition to the Native Breeds Support (NBS) list, which classifies breeds as "At risk", "Not at risk", or "Uncategorised" based on FAO thresholds for breeding females. Under the ELM scheme, farmers are paid to use these rarer native breeds in rotational and extensive grazing systems, recognising their ecological value in maintaining species-rich habitats. This integration helps sustain populations and embed conservation within productive farming systems. While the NBS improves targeting and accountability, it still focuses on breed identity rather than within-breed diversity. Defra should therefore complement it with an ex situ conservation initiative, as we do for seeds, and create a national livestock gene bank to store semen, embryos, and DNA from diverse lineages, providing long-term insurance against genetic erosion, disease, or climate-related loss.

Solutions for geopolitical resilience

The global food system has become an arena of geopolitical competition. Food is increasingly weaponised—through trade restrictions, cyberattacks, and strategic blockades—as states seek leverage in an unstable world. With around a quarter of the world's food crossing borders and nearly half of the UK's food imported, Britain's dependence on global markets leaves it acutely exposed.

Recent crises have shown how quickly politics on the other side of the world can interrupt supply in the UK: the war in Ukraine sent wheat and sunflower oil prices soaring; renewed US–China trade tensions have reshaped global commodity flows; and energy disruptions from the Iran conflict continue to ripple through fertiliser and transport costs. Geopolitical resilience must now sit at the heart of UK food policy. This requires moving from reactive crisis management to proactive security planning. Policy interventions should focus on diversifying supply sources and routes, rebuilding domestic production and processing capacity, and embedding food security within national-security strategy. Stronger international alliances, targeted investment in resilient infrastructure, and better coordination between government and industry will also be vital. By integrating these measures, the UK can transform vulnerability into preparedness—ensuring that its food system can withstand the political, economic, and technological shocks that increasingly define the 21st century. Actions that would deliver geopolitical resilience within the food system include.

Recognise food security as national security

In an era of growing geopolitical instability, food resilience must be recognised as a core element of national defence. Minister Eagle acknowledged this need at the launch of the APPG for Farming in October 2025 when she pronounced that "food security is national security". This echoed the Labour Party's Manifesto for the 2024 General Election.[19] Around the world, food is increasingly being weaponised—used as leverage in political and military conflict. Whether through targeted trade tariffs, export bans, or the deliberate destruction of agricultural infrastructure, food is becoming a strategic tool of war. As we discussed, the conflicts in Gaza, Iran and Ukraine starkly illustrated this. Hugely dependent on food imports, the UK is exposed to disruptions to the global food markets. The ripple effects of conflicts are substantial and risk destabilising food security. Yet, despite food's critical role in our national security, we do not treat it as we do other essential commodities. We stockpile oil, gas, and medical supplies to ensure resilience in times of crisis, but our food system remains dangerously exposed. The UK imports nearly half of what it eats, relying on long, complex global supply chains that are increasingly vulnerable to disruption.[20] To safeguard national stability, stable food access must be embedded within defence planning and risk management frameworks. Building resilient domestic supply chains, advancing greater domestic production of essential food, and diversifying trade routes to avoid choke points are not just agricultural policies—they are matters of national security. The UK Government must begin treating food security as a core component of national defence. Governments around the world are starting to recognise the strong links between food and conflict. Uganda, for instance, has merged its Ministry of Agriculture's National Agricultural Advisory

Services into its Ministry of Defence. Being simultaneously a driver and a consequence of conflict, as we have seen in many instances, including what is happening in Gaza right now, the World Food Programme agrees with the fact that food insecurity should not be tackled as a separate issue from that of national security.[21,22] Just as energy and health systems are protected as critical infrastructure, our food system (and the infrastructure that underpins it) must be safeguarded against future shocks.

By integrating food into national-security frameworks, the UK can move from a reactive to a preventative approach—one that ensures our citizens remain fed, stable, and secure even in times of crisis. Legislation should establish a statutory duty on the state to ensure the population can be fed during crises, following models such as Sweden, where national preparedness plans explicitly outline responsibilities for central and local government. This falls in line with the advice laid out in Professor Tim Lang's Resilience report, which was discussed at length earlier. Food security resilience must also be embedded into risk planning at every level from local to regional to national. The National and Community Risk Registers should incorporate detailed assessments of how shocks—such as supply chain breakdowns, trade disruptions, or extreme weather—could affect access to food, with greater consideration given to how communities would respond. Local authorities should be supported to conduct community food audits, mapping local assets, skills, and infrastructure that could help maintain food supplies during emergencies. Furthermore, the National Infrastructure Commission should formally include food within its remit, ensuring that its advice to government recognises the food system as an essential part of the UK's resilience architecture.

Consider cyber security of food supply chains

Yet, beyond geopolitical conflict disrupting our access to food, there is a risk of our food supply could be targeted explicitly—either by state or non-state actors. Cyberattacks add another dimension to this threat, with hackers targeting logistics systems, cold storage facilities, and even food safety databases. A successful attack on, for example, Tesco could cripple the UK's distribution of essential goods within days, undermining both supply and food safety. The UK food system remains vulnerable to intentional disruption—through intentional contamination or closure from cyberattacks. UK policy should move towards a more integrated, anticipatory model that links food defence, cyber resilience, and national infrastructure protection. First, the Government should establish a national food-defence coordination unit, under the FSA's National Food Crime Unit, modelled on the US FDA's Intentional Adulteration (IA) framework. This would help to standardise vulnerability assessments, mitigation strategies, and incident reporting across the sector. This would clarify roles between regulators, local authorities, and industry while ensuring that deliberate contamination risks are treated as national-security concerns rather than solely food safety issues. Second, the National Cyber Security Centre (NCSC) should develop sector-specific supply-chain cybersecurity standards for food and agriculture—adapting US CISA guidance—to address operational-technology (OT) and data-integrity risks in processing, logistics, and retail, and formalise an information-sharing and analysis mechanism, akin to the US Food and Ag-ISAC, enabling trusted, real-time exchange of threat intelligence among food companies, technology providers, and government agencies.

There are many actions across different levels of the food system that can strengthen the UK's food stability and build long-term resilience. At the farm level, individual farmers have an essential role to play. Even for those who are not ready or able to convert to organic certification, adopting more agroecological techniques—such as improving soil health, diversifying crops, and reducing reliance on synthetic inputs—can make farms more adaptable to climate and market shocks. Farmers can also explore alternative routes to market, such as local food shops, direct-to-consumer sales, and regional procurement contracts, which help retain value within rural communities. In addition, new income streams like biodiversity net gain schemes and soil carbon credits offer potential ways to reward farmers for environmental stewardship while supporting economic viability.

At the retail level, large and small food businesses must ensure that their environmental, social, and governance (ESG) commitments are more than token gestures but see them as risk management. Retailers have enormous power to shape consumer demand and influence the direction of farming through their sourcing choices. By curating product ranges that reflect the future of sustainable agriculture—prioritising local, low-impact, and fairly traded goods—they can drive systemic change throughout the supply chain. Retailers can also act as partners in transition, supporting farmers financially and contractually as they adopt more sustainable methods and adapt to changing market conditions. Central to this is creating a consumer-facing labelling system that could enable farmers to communicate their progress towards regenerative farming, inform the public, and help justify higher prices. Increasingly, UK consumers want to make ethical and environmentally responsible choices about the food they buy, particularly meat. Practices like extensive grazing, maintaining species-rich pastures, and high-welfare slaughtering tend to cost more, making transparent information essential for consumers to support sustainable producers. Currently, many shoppers encounter a variety of labels—Organic, Pasture for Life, Free Range—that can be confusing. There is growing recognition of the need for clearer, standardised method-of-production labelling, backed by independent verification, such as the Global Farm Metric, to provide consumers with accessible and reliable information at the point of purchase. Ultimately, clearer labelling represents an important tool in building a more resilient, ethical, and environmentally sustainable food system in the UK. Retailers leading this push and be at the forefront of providing transparent and comparative information to their customers.

Notes

1 GOV.UK, Spending review 2025 (policy paper), HM Treasury, 2025, https://www.gov.uk/government/publications/spending-review-2025-document/spending-review-2025-html
2 GOV.UK, Public sector food and catering policy for England. The Government Buying Standards for Food and Catering Services (proposed document for June 2022 consultation), DEFRA, 2022, https://consult.defra.gov.uk/public-sector-food-procurement/food-and-catering-consultation/supporting_documents/Proposed%20public%20sector%20food%20and%20catering%20policy%20for%20England%201.pdf
3 ReLondon, Hackney City Council & Sustain, London food purchasing commitment, 2023, https://relondon.gov.uk/wp-content/uploads/2025/10/London-Food-Purchasing-Commitment-toolkit.pdf
4 Simms, A., Britain's 'house of food' sits on shaky foundations. To fix it, curb the big supermarkets, *The Guardian*, 2023, https://www.theguardian.com/commentisfree/2023/aug/15/britain-food-big-supermarkets-profits-jobs-suppliers
5 McKee, A., et al., Increasing the availability of farmland for new entrants to agriculture in Scotland. Final report to the Scottish Land Commission, GOV.Scot, 2018, https://www.landcommission.gov.scot/downloads/5dd6a2d2ac866_McKee-et-al.-Final-report-to-SLC-Increasing-land-availability-for-new-entrants-2.5.2018.pdf
6 UKRI, Agri-tech and food technology in Eastern England, UKRI, 2024, https://iuk-business-connect.org.uk/opportunities/agri-tech-and-food-technology-in-eastern-england/
7 GOV UK, Small Holdings and Allotments Act 1908, UK Parliament, 1908, https://www.legislation.gov.uk/ukpga/Edw7/8/36
8 Graham, Kim. et al., Reviving county farms, CPRE, 2019, https://www.cpre.org.uk/wp-content/uploads/2019/12/December-2019_Reviving-county-farms.pdf
9 Leveson-Gower, H., New entrant support scheme: Apply for the pilot, DEFRA, 2022, https://defrafarming.blog.gov.uk/2022/11/24/new-entrant-support-scheme-apply-for-the-pilot/
10 GOV.UK, A UK government food strategy for England, considering the wider UK food system (policy paper), DEFRA, 2025, https://www.gov.uk/government/publications/a-uk-government-food-strategy-for-england/a-uk-government-food-strategy-for-england-considering-the-wider-uk-food-system
11 Mendez, S., Spain: Amendment of Law 12/2013 on measures to improve the functioning of the food supply chain (Law 16/2021 of December 14, 2021), *Global Compliance News*, 2022, https://www.globalcompliancenews.com/2022/01/23/spain-amendment-of-law-12–2013-on-measures-to-improve-the-functioning-of-the-food-supply-chain-law-16–2021-of-december-14–2021–13012022/
12 Dublin City Council, Dublin City Council launches refurbishment of the historic fruit and vegetable market, Dublin 7, 2025, https://www.dublincity.ie/news/dublin-city-council-launches-refurbishment-historic-fruit-and-vegetable-market-dublin-7
13 Loaiza Puerta, V., et al., Improvement of soil structure through organic crop management, conservation tillage and grass-clover ley. *Soil and Tillage Research, 180*, 1–9, 2018. DOI:10.1016/j.still.2018.02.007

14 Soil Association, Organic standards for Great Britain, Farming and growing, 2025, https://www.soilassociation.org/media/23378/gb-farming-growing.pdf
15 Soil Association, Certification fees for farming & growing to Soil Association standards or organic regulations (in GB or NI), 2025, https://www.soilassociation.org/media/28599/fees-for-farming-and-growing-to-sa-standards-or-organic-regulations-in-gb-or-ni-2025–26.pdf
16 Carlile, C. & O'Donnell, E., Staff shortages have forced major cuts to Environment Agency water quality monitoring, internal documents reveal, *Unearthed, Greenpeace UK*, 2025, https://unearthed.greenpeace.org/2025/07/24/environment-agency-cuts-water-monitoring/
17 DEFRA & UKRI, Farming innovation: Find out about funding, UKRI, 2021, https://farminginnovation.ukri.org
18 Global Partnership for Plant Conservation, *The global strategy for plant conservation 2020–2030*. Botanic Gardens Conservation International (BGCI), 2025.
19 The Labour Party, Supporting British farmers. In: *Labour Party Manifesto 2024*. 2024, https://labour.org.uk/change/make-britain-a-clean-energy-superpower/#farmers
20 GOV.UK, UK food supply sources. In: *United Kingdom Food Security Report 2021*. DEFRA, 2022, https://www.gov.uk/government/statistics/united-kingdom-food-security-report-2021/united-kingdom-food-security-report-2021-theme-2-uk-food-supply-sources
21 World Food Programme USA, Should food security be a national security issue? WFP USA, 2019, https://wfpusa.org/news/should-food-security-be-a-national-security-issue/#:~:text=Putting%20food%20security%20on%20the,responsibility%20to%20feed%20our%20citizens
22 UN, "Weaponization of food in Gaza" as over 400 Palestinians have been killed seeking aid, OHCHR, 2025, https://www.ohchr.org/en/press-briefing-notes/2025/06/gaza-palestinians-seeking-food-continue-be-killed-israeli-military

Conclusion

Policy for the future of food and farming

The UK's food system is facing unprecedented strain, having historically prioritised efficiency and yield. It has become dangerously reliant on just-in-time logistics and concentrated corporate supply chains—dependent on global imports to guarantee food security. Supply chain disruptions have pushed food prices to historic highs. Food price inflation has hit 45-year highs, geopolitical events driving up the costs of essential inputs, such as energy and fertiliser, and increasingly restricted labour markets. Emerging global trends underscore the urgency of this shift. The war in Ukraine weaponised wheat and fertiliser supplies, while renewed trade wars, particularly between the US and China, have disrupted global commodity markets. A pattern of high volatility and weakening stability in the UK has emerged, revealing the deeply interconnected nature of vulnerabilities in the global food system. Climate change, biodiversity loss, and water insecurity remain urgent background threats. Climate change is accelerating. In July 2024, The Guardian reported that UK farm incomes fell by 19% due to widespread flooding that left crops unable to be harvested.[1] Such events will only become more frequent and severe. Without intervention, the fragility of the food system will only deepen. Over the past decade, acute shocks and chronic pressures have been colliding and compounding in ways that destabilise our food system.

Not just a theoretical problem for the policy-minded, this is having a damaging impact on citizens—food insecurity is rising despite increases in per capita food production. By 2023, the UK recorded the highest level of food price inflation in 45 years, outpacing general consumer inflation and ranking among the highest in the G7, showing that the problem is not only about supply but also about access, equity, and systemic vulnerability.[2] The combination of rising food and energy prices squeezed household budgets, forcing many consumers to prioritise price over health or sustainability. For some, this meant cutting back on fruit, vegetables, and higher-quality proteins; for others, it meant skipping meals altogether. The result has been rising food insecurity and growing inequalities in diet and health outcomes. Vulnerable groups—low-income families, households with children, and people with disabilities—have been hit hardest.

DOI: 10.4324/9781003581871-6

To tackle this dynamic, the UK Government has made attempts to address the issue of food security, but these attempts have been half-hearted and siloed. They have tinkered around the edges with commissions, white papers and strategies—never really giving the issue the attention and resourcing it deserves. Consequently, they have failed to grapple with the enormity of the problem. As leading food thinker, Pen Vogler, observes, "there is a lack of knowledge or curiosity about the interconnectivity of good food and a good society and the recognition that both require long-term planning".[3] This is precisely what must change. The path forward requires courage and imagination to develop a concerted focus on food resilience. We need circular, regenerative approaches that work with nature, not against it—systems that value biodiversity, rebuild soil health, and support communities. We must also rethink how we define success: not by short-term yields or profits alone, but by long-term stability, equity, and ecological balance. This needs to be an active process. We cannot just cross our fingers and hope that it will all be okay, and another disruption will not cripple our food supply.

But what is food resilience? The term itself is difficult to define; without an agreed definition, it is impossible for the mechanisms of the state to function to deliver it. It refers to the capacity of systems—ecological, social, and economic—to withstand, adapt to, and recover from disruption while continuing to provide safe, nutritious, and affordable food. It bridges food security (ensuring availability and access) with food sovereignty (empowering communities to shape their own food systems), recognising that resilience depends on both supply stability and local agency. At its core, resilience is dynamic—it is not just about "bouncing back" after shocks such as conflict, climate extremes, or pandemics, but about moving forward, evolving to reduce future vulnerability. The best definition is from the FAO that encompasses these complexities and defines resilience as "the ability of agroecosystems, communities, and food systems to absorb shocks and stresses—whether environmental, economic, or social—and continue to function and evolve. It means bouncing back from disruptions while adapting and transforming over time".[4] Agroecological approaches exemplify this, using diversity and circularity to buffer against risk.

Resilience cannot merely be hoped for—it has to be actively built into our food system with intent and purpose. In an increasingly uncertain world—marked by climate shocks, geopolitical instability, supply chain vulnerabilities, and economic inequality—resilience is no longer a luxury; it is a necessity. In this increasingly chaotic world, resilience matters across all sectors, from health and energy to infrastructure and finance. But in food systems, it is uniquely urgent. Disruption to food access or supply has immediate social and political consequences. The OECD acknowledges that weathering shocks to our food system needs consideration and planning, saying, "a proactive approach focused on building resilience requires preparedness and reconciling the abilities to absorb, adapt and transform".[5] Professor Tim Lang reflected this urgency and proactive necessity in his recent report.[6] Moreover, they need to be funded. Positive words cannot drive the level of change that is needed. Concretely, building resilience into our food system means segmenting the different aspects within food resilience so as to ensure focus and attention are given to each sub-category equally. Building resilient food systems means designing them to endure volatility—balancing productivity, equity, and ecological integrity—so that both producers and consumers can thrive through uncertainty and change.

The best opportunity to ensure a holistic approach to food resilience is adopted by the Government is in the forthcoming Action Plan for the 2025 Food Strategy. The first step to grasping the opportunity to build food resilience into the Action Plan is to consider farming and food together, recognising the entire system from field to fork. Such an integrated approach is the only way to deliver a resilient and future-facing food framework for the next generation. UK food and farming policy has historically been treated as separate, yet they are inextricably linked and must be integrated to deliver real change. Farming determines not only what food is produced, but also how healthy, affordable, and sustainable that food is. When agricultural policy focuses narrowly on what happens on the farm and in the field, it can drive overproduction, environmental damage, and diets high in calories but low in nutrition. Conversely, food policy that only addresses consumption misses the structural conditions that shape what is grown, who grows it, and how it reaches people. By truly integrating food and farming policy, governments can align incentives for farmers with public health and climate goals—supporting the production of fruit, vegetables, and pulses that underpin healthy diets while restoring soils and biodiversity. This joined-up approach ensures that taxpayer support for farming directly contributes to healthier diets, lower emissions, and resilient local economies, rather than working at cross-purposes. The next step to developing true food resilience is to expand the understanding to encompass all six sub-categories. The Government's recognition of the importance of resilience in supply chains in the National Food Strategy is certainly welcome.[7] However, that is only one aspect of resilience. Resilience is more than just ensuring supply chains can recover from shocks. To make our food system resilient, all aspects of resilience need to be actively considered. Resilience in food systems operates across the discussed subcategories: climate, genetic, economic, market, distribution, and geopolitical. Ultimately, doing so would further the Government's aims, since a broader understanding of food resilience will further deliver against supply chain resilience too, since all aspects are connected and one cannot be achieved without the other.

The approach of segmenting off the different aspects of food resilience, tackling each with equal weight by recognising their interconnectivity, is a new approach. Consequently, it deserves a new terminology: 'food stability'.

Food stability is the ability of the food system to withstand disruption and potential threats by using resilience and diversity to mitigate risk. Food stability goes beyond short-term emergency planning or stockpiling. It's about designing a system that can adapt to shocks, absorb disruption, and recover quickly, all while ensuring equitable access to nutritious, affordable food. By giving food resilience its own conceptual space—through the term 'food stability'—we can better target policy, funding, and innovation towards long-term security. It also helps break down the silos between health, environment, trade, and agriculture, offering a more joined-up approach to risk. A dedicated focus on food stability allows us to be proactive rather than reactive, ensuring the food system can withstand whatever challenges lie ahead.

To achieve food stability, we need to actively build resilience into our food system. For those who back 'food sovereignty', agroecology has proven itself to be the answer to the need for increased resilience. It is true that agroecological systems are more resilient since they have a greater capacity to recover from drought, floods, or hurricanes.[8] Furthermore, they foster stronger socio-economic resilience because by diversifying the range of crops grown and reducing dependence on external inputs, producers dilute economic risk. Plus, greater domestic production means less reliance on imports. Yet, systemic stability is not their goal. Resilience is not their raison d'etre. It is a byproduct of a farming system that focuses on natural systems. Similarly, the food security side might argue that they work to guarantee that there are sufficient calories for all. Yet, once again, they are not concerned with the stability of future supply. They are concerned with the here and now. That is important, but we need to keep an eye on the horizon to try to pre-empt what might be coming and work to mitigate those risks. We need a food policy that is focused on delivering resilience to avoid future disruption.

By explicitly adding the concept of food stability, UK policy can create a new dedicated focus on an integrated view of food resilience. Food stability recognises that acute shocks—whether extreme weather, pandemics, wars, or cyberattacks—will continue to disrupt the system. It requires governance that anticipates, mitigates, and adapts to such crises. The EU's Common Agricultural Policy is structured around two pillars, but the UK post-Brexit must look to the future and adopt a tripartite framework—food security, food sovereignty, and food stability.[9] A Food Security Pillar would ensure that hunger is eradicated and access to affordable food is guaranteed, a Food Sovereignty Pillar would restore ecosystems, support farmers, and embed sustainability, and a Food Stability Pillar would hardwire resilience into the system, focusing on crisis preparedness, diversified supply chains, and adaptive governance. Crucially, these pillars must be weighted equally. A three-legged stool is stable only if all legs are sound; weaken one, and the whole system collapses. Unlike the Common Agricultural Policy model, which became bureaucratic and lacked aspiration, the UK's framework must be visionary and dynamic.

For decades, food and farming policy stagnated, constrained by political reluctance to challenge entrenched subsidy systems. Brexit, though disruptive, offers an opportunity to build something new.

If the Government were ambitious it would introduce a commitment to deliver a Food Bill that reflected this three-pillar approach. Such legislation could be a once-in-a-generation opportunity to address systemic failures of our current food system and create a UK food future that is healthy, equitable, and sustainable. A similar opportunity to be bold was presented by the 2020 Agriculture Act, but that was sadly squandered. This time, the opportunity needs to be grasped, and not squandered. A visionary Food Bill could set a leading standard for other Global North nations, showing how to align productivity, environmental stewardship, and social justice within a coherent national strategy. The last Food Act was the Food Safety Act in 1990, which strengthened consumer protection and improved alignment with EU food standards. It essentially focused on hygiene and safety.[10] A Food Act would need to take a much more holistic approach to food to create a food system that is healthy, equitable, and sustainable. It is critical that this process considers both food *and* farming and looks at the entire food system as a whole. It would be an error to only consider the food side (i.e. once the food has left the farm and becomes a consumer good within a supply chain). Such a commitment would demonstrate a true intention to tackle the fundamental flaws of our food system today and create one that is future-facing and fully recognises and tackles the challenges that our food system is currently facing.

In conclusion, the UK stands at a crossroads. We can continue to patch cracks in a fragile system, lurching from crisis to crisis, or we can build a resilient food future. The case for resilience is not abstract: it is written in flooded fields, in empty supermarket shelves, and in the rising tide of food insecurity among families. Resilience will require public investment. Climate adaptation—whether through water storage, soil restoration, or flood defences—cannot be funded by farmers or consumers alone. The Government must recognise food resilience as a matter of national defence and fund it accordingly. As climate shocks intensify and global trade becomes more precarious, the UK needs to build buffers, redundancy, and flexibility into its food infrastructure to recognise the interconnectivity of food and a healthy society. That must change. Food is not merely an economic commodity—it is the foundation of health, community, and social stability. Investing in food resilience is therefore an investment in the country's long-term security. To adequately deliver food resilience, the concept of food stability could provide a helpful focus. Food stability recognises that shocks—pandemics, wars, floods, cyberattacks—will continue to occur and must be planned for. A new Food Bill, rooted in food security, sovereignty, and stability, offers the opportunity to break from decades of inertia and to craft a visionary policy that serves citizens, farmers, and the planet. Food Security would guarantee access and affordability; Food Sovereignty would restore ecosystems and empower producers; and Food Stability would focus on crisis preparedness, diversified supply chains, and adaptive governance. The UK should seize the opportunity to enshrine this vision in a new Food Bill, modelled on the EU's multi-pillar approach but with greater ambition and dynamism, but this will require investment, regulation, and international cooperation. But more than that, it will require political will—the determination to see food not just as a market commodity, but as critical infrastructure and a public good. If the UK can take this step, it can create a food system capable of withstanding the storms of the 21st century and ensuring that every citizen has secure, sustainable, and stable access to good food. A visionary Food Bill, rooted in security, sovereignty, and stability, offers the chance to secure the nation's food future. It demands political courage, but the reward—a resilient, fair, and sustainable food system capable of withstanding the shocks of the 21st century—is one worth striving for.

Notes

1 Horton, Helen, Floods fuelled 19% drop in income from farming in England in 2023, *Guardian*, 2024, https://www.theguardian.com/environment/article/2024/jul/12/floods-fuelled-19-drop-in-income-from-farming-in-england-in-2023

2 UK.Gov, Food and energy price inflation, UK: 2023, Office of National Statistics, 2023, https://www.ons.gov.uk/economy/inflationandpriceindices/articles/foodandenergypriceinflationuk/2023#:~:text=imported%20food%20prices-,UK%20food%20price%20inflation%20was%20among%20the%20highest%20across%20G7,Notes:

3 Vogler, Pen, *Stuffed: A history of good food and hard times in Britain*. Atlantic Books, 2024.

4 FAO, Resilience: Enhanced resilience of people community and ecosystems is key to sustainable food and agricultural systems, Agroecology Knowledge Hub, https://www.fao.org/agroecology/knowledge/10-elements/balance/en/?page=5&ipp=5&tx_dynalist_pi1%5Bpar%5D=YToxOntzOjE6IkwiO3M6MToiNSI7fQ%3D%3D

5 OECD, Resilience in agriculture and food systems, https://www.oecd.org/en/topics/sub-issues/resilience-in-agriculture-and-food-systems.html

6 Lang, Tim, et al., Just in case: Narrowing the UK civil food resilience gap, National Preparedness Commission, 2025, https://nationalpreparednesscommission.uk/wp-content/uploads/2025/02/NPC-Just-in-Case-Main-Report_PDF-Download.pdf

7 UK.Gov, Policy paper: A UK government food strategy for England, considering the wider UK food system, DEFRA, 2025, https://www.gov.uk/government/publications/a-uk-government-food-strategy-for-england/a-uk-government-food-strategy-for-england-considering-the-wider-uk-food-system

8 FAO, Resilience: Enhanced resilience of people community and ecosystems is key to sustainable food and agricultural systems, Agroecology Knowledge Hub, https://www.fao.org/agroecology/knowledge/10-elements/balance/en/?page=5&ipp=5&tx_dynalist_pi1%5Bpar%5D=YToxOntzOjE6IkwiO3M6MToiNSI7fQ%3D%3D
9 Europea.eu, The common agricultural policy: An overview, EU CAP Network, https://eu-cap-network.ec.europa.eu/common-agricultural-policy-overview_en
10 Gov.uk, Food Safety Act 1990, Legislation, https://www.legislation.gov.uk/ukpga/1990/16/section/60?

Recommended further reading

Altieri, M.A. (1995) *Agroecology: The science of sustainable agriculture*. Boulder, CO: Westview Press.

Brown, G. (2018) *Dirt to soil: One family's journey into regenerative agriculture*. White River Junction, VT: Chelsea Green Publishing.

Chatham House (2020) *Food system impacts of trade and geopolitics*. London: Chatham House.

Cribb, J. (2010) *The coming famine: The global food crisis and what we can do to avoid it*. Berkeley, CA: University of California Press.

De Schutter, O. (2017) *The political economy of food systems reform*. New York: UN Special Rapporteur on the Right to Food.

Dimbleby, H. (2020) *Ravenous: How to get ourselves and our planet into shape*. London: Profile Books.

FAO (2018) *The 10 elements of agroecology*. Rome: Food and Agriculture Organization of the United Nations.

FAO (2023) *The state of food security and nutrition in the world*. Rome: Food and Agriculture Organization of the United Nations.

Gaffney, O. and Rockström, J. (2021) *Breaking boundaries: The science of our planet*. London: Penguin Random House.

Kolbert, E. (2014) *The sixth extinction: An unnatural history*. New York: Henry Holt and Company.

Lang, T. (2020) *Feeding Britain: Our food problems and how to fix them*. London: Pelican.

Levinson, M. (2006) *The box: How the shipping container made the world smaller and the world economy bigger*. Princeton, NJ: Princeton University Press.

Marmot, M. (2015) *The health gap: The challenge of an unequal world*. London: Bloomsbury.

Montgomery, D.R. (2017) *Growing a revolution: Bringing our soil back to life*. New York: W.W. Norton & Company.

Nestle, M. (2013) *Food politics: How the food industry influences nutrition and health*. Berkeley, CA: University of California Press.

Patel, R. and Moore, J.W. (2017) *A history of the world in seven cheap things: A guide to capitalism, nature, and the future of the planet*. Oakland, CA: University of California Press.

Pollan, M. (2006) *The Omnivore's dilemma: A natural history of four meals*. New York: Penguin Press.

Rodrik, D. (2011) *The globalization paradox: Democracy and the future of the world economy*. New York: W.W. Norton & Company.

Sassen, S. (2014) *Expulsions: Brutality and complexity in the global economy*. Cambridge, MA: Harvard University Press.

Smaje, C. (2020) *A small farm future*. White River Junction, VT: Chelsea Green Publishing.

Tree, I. (2018) *Wilding: The return of nature to a British farm*. London: Picador.

Wilson, B. (2019) *The way we eat now*. London: Fourth Estate.

Acronyms

AAFC	Agriculture and Agri-Food Canada
AHDB	Agriculture and Horticulture Development Board
AI	Artificial intelligence
APPG	All-Party Parliamentary Group
AUD	Australian dollar
BASF	Badische Anilin- und Soda-Fabrik
BCM	Business Continuity Management
BEIS	Department for Business, Energy, and Industrial Strategy
BMEL	German Federal Ministry of Food and Agriculture
BSI	British Standards Institution
CAP	Common Agricultural Policy
DEFRA/Defra	Department for Environment, Food, and Rural Affairs
DLUHC	Department for Levelling Up, Housing, and Communities
DTI/DST	Department for Science, Innovation, and Technology
FAO	Food and Agriculture Organization of the United Nations
FSA	Food Standards Agency
FSS	Food Standards Scotland
FTA	Free Trade Agreement
GBS	Government Buying Standards
GCA	Groceries Code Adjudicator
GHG	Greenhouse gas
HFSS	High fat, salt, and sugar (foods)
IA	Impact assessment
IPCC	Intergovernmental Panel on Climate Change
ISAC	Information Sharing and Analysis Centre
ISDS	Investor-State Dispute Settlement
LGA	Local Government Association
LRF	Local Resilience Forum
MEP	Member of the European Parliament
MLA	Member of the Legislative Assembly
NAFTA	North American Free Trade Agreement
NBS	Nature-based solutions
NPC	National Preparedness Commission
OBR	Office for Budget Responsibility
OECD	Organisation for Economic Co-operation and Development

PBO	Precision-bred organism
PHE	Public Health England
R&D	Research and development
SDS	Sustainable Development Strategy
SFI	Sustainable Farming Incentive
SME	Small- and medium-sized enterprise
UN	United Nations
UNCCD	United Nations Convention to Combat Desertification
UNFCCC	United Nations Framework Convention on Climate Change
USDA	United States Department of Agriculture
WFP	World Food Programme
WHO	World Health Organization
WWF	World Wide Fund for Nature

Index

For Product Safety Concerns and Information please contact our EU representative GPSR@taylorandfrancis.com Taylor & Francis Verlag GmbH, Kaufingerstraße 24, 80331 München, Germany